CREW RESOURCE MANAGEMENT

FOR THE FIRE SERVICE

CREW RESOURCE
MANAGEMENT
FOR THE FIRE SERVICE

Randy Okray
and
Thomas Lubnau, II

Copyright© 2004 by
PennWell Corporation
1421 South Sheridan Road
Tulsa, Oklahoma 74112-6600 USA

800.752.9764
+1.918.831.9421
sales@pennwell.com
www.FireEngineeringBooks.com
www.pennwellbooks.com
www.pennwell.com

Marketing Manager: Julie Simmons
National Account Executive: Francie Halcomb

Director: Mary McGee
Production / Operations Manager: Traci Huntsman
Managing Editor: Jared d'orr Wicklund
Production Editor: Sue Rhodes Dodd
Book Designer: Robin Remaley
Cover Designer: Matt Berkenbile

Library of Congress Cataloging-in-Publication Data Available on Request

Crew Recource Management for the Fire Service / by Randy Okray and Thomas Lubnau, II
    p. cm
Includes index.
ISBN 978-1-59370-006-5

Printed in the United States of America

6 7 8 9 10   16 15 14 13 12

A special thank you to my wife who supports me—and all my wild ideas.
To my children, Austin and Brandon,
I hope that you find a safer world in your life.

—*Randy Okray*

To Rita, Rachel, and Tommy, thank you for your patience and your love.
I like to believe that what we have done here will save lives.
It could never have been accomplished without your help.

—*Tom Lubnau*

To all of our friends with numbers on the sides of your helmets,
may you go home safely to your family tonight.

—*Randy and Tom*

FOREWORD

CICERO SAID, "TO ERR IS HUMAN." More than 2000 years later, that simple statement has become the mantra for a concept that is revolutionizing the way mistakes, injuries, and deaths are being prevented in the aviation, medical, and military industries. The concept, crew resource management (CRM), emphasizes that in most cases technology isn't the root cause of human tragedy. More often *human* error is the root cause of catastrophe. Therefore steps taken to minimize the impact of errors are the keys to reducing the magnitude of the inevitable human failing.

Fire service history is replete with examples of the effects of human error. Hackensack, Storm King Mountain, Washington, D.C., and Lairdsville, NY, conjure up images of firefighters performing their duties and suffering horrible consequences. Fire service equipment, standards, education levels, and knowledge have advanced a firefighter's ability to extinguish fires exponentially over the last 25 years. However, the death and injury rate for firefighters plateaued approximately 10 years ago and continues to average 95 fatalities and 95,000 injuries per year. Today's firefighters are better protected, better trained, and a smarter lot than their predecessors, yet the death rate has remained static. CRM offers a new approach to reduce those deaths and injuries.

No one intentionally sets out to make a mistake. Typically an accident is the result of a series of events that form a chain from initiation to catastrophe. Employing CRM has proven to be an effective method to break the links of the error chain, thereby minimizing or completely averting disaster.

The work you are about to read may be your first step toward a proven method for preventing tragedy and failure or a continuation of a journey to reduce the effects of human error. Regardless of your standing in the error management field, you are to be commended for taking up the challenge. Tom Lubnau and Randy Okray have been studying, practicing, and living CRM for nearly 10 years. The members of the Campbell County Fire Department, Gillette, Wyoming have proven that CRM can be applied successfully to the fire service. You will find this work written in a straightforward style that will enlighten and equip you with the knowledge you can use to begin the greatest challenge known to humankind—preventing human error.

JOHN TIPPETT
DISTRICT CHIEF, MONTGOMERY COUNTY (MD)
DIVISION OF FIRE AND RESCUE SERVICE,
ASSISTANT TASK FORCE LEADER AND
SAFETY OFFICER FOR MARYLAND TASK FORCE 1
IAFC IN SPECIAL PROJECTS

For more information regarding Crew Resource Management, including training opportunities and resources, contact the authors at:

randyo@vcn.com
tlubnau@collinscom.net

INTRODUCTION

In 1996, I stood on a steep mountainside in Colorado with a fellow firefighter. We stood next to a white stone cross, our chests heaving and sweat pouring from our faces. From all known reports and estimates, we had attempted to replicate the "run for life" that consumed the last minutes of 14 firefighters' lives. We carried a light day pack, much lighter than the 40+ pound line packs that the firefighters carried. We had no tools to slow us down; we had not been working all night and day under excruciating environmental and physical conditions. Yet there we stood, at the first cross—hundreds of feet downhill from the ridge top and the shielded safety the other side had offered many of our fellow firefighters, the survivors, just two years ago.

That's when it hit me. I could have died at this cross. My name could also be chiseled in this peculiar stone. How did I know that? I knew from the events of that day and from the actions that Don Mackey, a smoke jumper, took that awful day in July 1996. He knew that the firefighters were in trouble—some say he had a "gut feeling" that things were not going well. He knew that everyone must get out—*now!* He left his fellow crew members, who had started their escape, to warn, assist, and direct firefighters who he probably had just met not more than a day or two earlier; but, with whom he had a fraternal bond. I would not have hesitated to do what Don had done.

Unfortunately, for the 14 firefighters who died that day and also those who survived, the cards had already been dealt. They were stuck with the hand they had—some won, some lost. Don Mackey, a firefighter's firefighter, died that day while saving firefighters. He could have made it to safety at least two different ways. He could have stayed with his own crew and let the others fend for themselves, but he was not the type of leader who could do that.

When the firefighters he was running with began to slow with stinging pains in their chests and cramping legs, he could have easily passed them. Don was in excellent physical condition. But Don chose to help those firefighters, push them on toward the ridge top, and shout at them to, "Keep going! We're almost home! Don't give up." When I reflect upon those incidents, I can almost hear his voice as I sit next to that cross.

Along with the feelings I had that day, and ever since then, I move forward with the concepts I learned in a paper by Ted Putnam, Ph.D., "The Collapse of Decision-Making and Organizational Structure on Storm King Mountain."

As for my writer colleague, Tom Lubnau, a roof collapsed on him and his partner in a structure fire. It almost ended his firefighting career and could have ended him. But from that day forward, the implementation of this program in our department and nationally became his priority too.

This book is our work in progress.

—RANDY OKRAY

CONTENTS

**1 Crew Resource Management (CRM):
Purpose of this Book** 1

The Problem with Safety Programs 5
Crew Resource Management 10
The Cards We Are Dealt 19
Overview of Chapters 21
References 23

2 Organizational Safety Culture 25

Organizational Culture Change—Not More Programs 27
The Error-free World 31
The Individual Responsibility 32
Barriers to Implementation 33
Creating the Safety Culture 36
References 38

3 Mission Analysis and Planning 39

The Formula 41
Micro-Training Opportunities 48
Positive, Proactive Attitude 50
Accountability 53
The Risk of Operational Activity 53
Risk-Versus-Gain Analysis 58
Risk Acceptance 58
The Changing Operational Setting 62
References 64

4   Situational Awareness **65**

What Is Situational Awareness? 65
How Do We Train for Situational Awareness? 66
When Do We Lose Situational Awareness? 68
Clues to Loss of Situational Awareness 69
Developing Tools to Maintain Situational Awareness 84
A Strategy for Maintaining Situational Awareness 86
Memory 90
Strategies for Increasing Memory .91
Conclusion 94
References 95

5   Communications **97**

Types of Communications 98
What if Computers Operated Like People? 103
Why Are You Communicating? 104
Orders Are Made To Be Followed 107
Guilty as Charged—Assumptions That Could Kill You 108
I Am Aware of the Problems—What Now? 112
The Communication System 114
Filters 124
So, What is Effective Communication? 128
References 130

6   It's CRM Leadership **133**

The First Step—Introduction 138
Second Step—Integration 139
Third Step—Trust 141
Who Is the Leader? 143
What Should a Leader be Doing? 150
Too Much, Too Late 151
Why Worry About Building a Team? 153
Team Performance Issues 153
Building the Team 156
The Organization's Role in Team-building 161

Teams: The Leader's Responsibilities 165
Practical Leadership 177
Where the Rubber Meets the Road 180
Conclusion 187
References 187

7 Followership **189**
What Is Followership? 190
Tendencies of Junior Personnel 191
Hurry-Up Syndrome 201
Recommendations 203
Conclusion 224
References 225

8 Decision-Making **229**
Decision Models 230
There Is a Substitute for Experience 233
Toward Better Decision-making 234
Decision-Making Aids 242
The Decision-Making Environment 246
Keys to Good Decision-Making 247
On-Scene Tips 249
Conclusion 250
References 250

9 Debriefing and Critiques **253**
The Missed Opportunities 255
Debriefs, Critiques, After-Action Reviews,
    and Other What-cha-ma-call-its 255
The Atmosphere is "The Blue Line" 257
The Pathway to Enlightenment 258
The Dos and the Do Nots 263
SFRM 265
The Wrap-up 267
References 270

10 Strategies for Implementation 271

The Training Effort 274
Conclusion 278
References 278

Index 279

CREW RESOURCE MANAGEMENT (CRM): PURPOSE OF THIS BOOK

In the 10 years it will take CRM to be introduced nationally, we will attend 1000 firefighter funerals... I can't get that out of my mind.

—GARRY BRIESE,
EXECUTIVE DIRECTOR,
INTERNATIONAL ASSOCIATION OF FIRE CHIEFS

OUR PURPOSE FOR WRITING this book is to transfer concepts that we learned from other industries (e.g., airline industry) to the fire service—to you. These concepts have millions of dollars of research and development behind them. They have years of hard work dedicated to their success. Also, they have proven themselves to save the lives of those who use them, and to save the lives of those individuals who place their lives in the hands of the strangers who use the concepts. Our purpose is to maybe, someday, give you something that will save your life, the lives of your fellow firefighters, and the strangers you serve and who trust you wholly.

CRM was first implemented officially at United Airlines, in the late 1970s; they called it CLR (Command, Leadership, Resource Management). Various aviation accidents highlighted the need to begin focusing on the major cause of aircraft disasters—human beings! Aircraft had become very reliable; yet, they continued to crash (although at a much lower rate than they had previously). Why did they continue to crash if the airplanes themselves were not breaking as much? Human beings were making errors.

The situations that a flight crew encounters are very complex and very stressful. Sometimes they made errors in perceiving the situation; sometimes they made errors in decisions; and sometimes, they made errors in their actions. The commercial aviation industry needed to fix this problem. So, with the help of the National Transportation Safety Board (NTSB), the Federal Aviation Administration (FAA), the National Aeronautics and Space Administration (NASA), and many private and public researchers, the study of human factors in aviation took off.

In 1978, a United Airlines DC-8 was inbound for Portland International Airport. The pilot, first officer, and the flight engineer as well as eight flight attendants and 181 passengers had an uneventful flight until the landing preparations began. The landing gear indicator lights showed that the nose gear was not down. The flight crew began a normal procedure for this failure, which entails breaking off the landing approach with instructions from air traffic control to proceed to an area where they could circle and troubleshoot the problem.

After running through the checklists, the flight crew saw that the nose gear still did not indicate that it was down properly. While circling, the aircraft had become low on fuel. The first officer and the flight engineer informed the pilot on various occasions of the situation. Probably due to the concentration on an abnormal issue and task, lack of delegating skills, communications issues, and maybe many others, the pilot did not realize he was placing his aircraft under a state of emergency.

The plane simply ran out of fuel and crashed six miles from the airport in a residential area. Ten people were killed and 23 were seriously injured. Luckily, the lack of fuel prevented a significant fire, which would surely have killed many, many more.

Why is this crash so significant? The cause of all this was a simple light bulb in the landing gear indicator. Millions of dollars in damages, many deaths, serious injuries, and untold lifelong pain and suffering caused by a light bulb that was probably worth half a dollar. Or, were there issues beyond the light bulb? The NTSB's investigation of the incident said *yes*. They began to look at incidents and accidents in a new fashion and they began to research how those incidents of minor mechanical failure could be stopped from becoming a human factors disaster.

What the commercial aviation industry discovered was that this stuff actually works. They decreased their accident rates. The American Fire Service is close to where the Commercial Aviation Industry was prior to their realization of the problem and implementation of Human Factors Training Programs. The fire service has improved their equipment and that equipment is reliable and easy to use. We spend millions of dollars on training to use this equipment and researching fire itself and understanding it. We train firefighters in the proper methods for fighting fires. We train them on what a fire does to a building or a hillside. We give them access to a variety of scientific or research-based resources to more fully understand their area of operations. But we are still killing about 100 firefighters a year.

Improvements in protective clothing, self-contained breathing apparatus (SCBA), fire knowledge, additional equipment, improved techniques, etc. do not seem to affect this benchmark of 100 firefighters lost. Why? Why doesn't the improved bunker gear save lives? Why won't better equipment save lives? Because, most of what kills our firefighters are human issues—not technical issues. How much time and money is spent, in your department, for training the physical side of your job? How much time and money is spent training the mental side of your job? Also, how much time and money is spent training for teamwork? We need to split some of the time and money off to fill in the voids in our training plans to address the mental and teamwork side of training.

Current statistics show a gradual 17% decrease in firefighter deaths between 1989 and 1998. Statistics from the same time period show about the same 17% decrease in total firefighter injuries. This is good news. This means that all the hard work, education, training, and equipment improvements are having an effect. But at about 85,000 to 100,000 injuries a year and around 100 firefighter deaths a year, this is still not an encouraging trend (see Fig. 1–1).

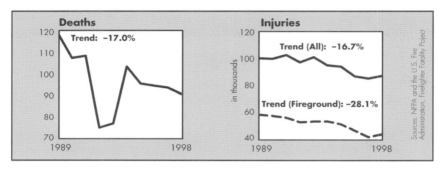

Fig. 1–1 Trends in Firefighter Casualties

Although these trends may appear to show that we are slowly addressing firefighter deaths and injuries, we must look at the total picture to get a better idea of what is happening. Data from the United States Fire Administration (USFA) for the time period, 1989–1998, shows that the trend for the number of fires nationally has decreased 13%, and the trend for civilian fire deaths and injuries have both decreased about 20%. This means that although our firefighters' deaths and injuries are decreasing, so are the number of fires and the severity of fires. This, of course, does not address the increased number of responses due to other duties such as emergency medical services (EMS), Technical Rescue, etc. Conversely, the statistics do not reflect the number of accidents, incidents, and property damage cases that firefighters are involved in during the course of their duties (see Fig. 1–2).

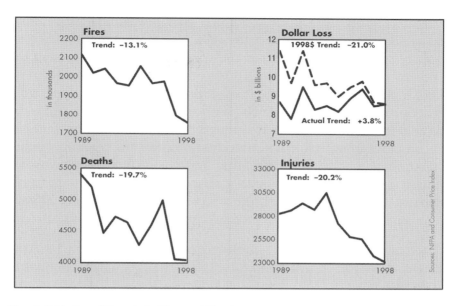

Fig. 1–2 National Trends in Fire and Fire Losses

THE PROBLEM WITH SAFETY PROGRAMS

Normal safety programs deal too much with technical-based solutions. They teach us how to do our jobs safer. They establish officer positions to monitor how we do our jobs. Also, they shoot safety slogans at us like BBs out of a shotgun. The problem with this approach is that it is only a portion of the problem. Effective safety programs not only need to address the technical and organizational sides of our jobs, but also the behavioral side—the human side.

We have all been told in our new "safety culture" that "everyone is a safety officer," and anyone who sees an unsafe operation has the authority to stop that action. The problem with this is that safety is often subjective and not objective. Seeing a firefighter inside a structure fire without an SCBA is a fairly objective item that is easily identifiable and defendable. Everyone on the fireground should know that this is not an acceptable practice. An interior attack on a fire burning in a structure for 15 minutes is less tangible and harder to stop for safety reasons.

Many "safety infractions" are based on a firefighter's perceptions and as such are sometimes very personal, a gut feeling even. These are hard to explain and justify, and the consequences of continuing the action or stopping the action are often unclear. It is often a perception of risk more than an actual violation of Standard Operating Procedures (SOPs) or appropriate work practices that turns out to be the contentious item on the fireground. Later you will learn that our subconscious mind traps more information than our conscious mind. Therefore, sometimes we *just know* and we do not know why we know it and cannot explain it to someone else, easily, why we know what we know. These gut feelings can often be a firefighter's savior when all other tools fail him.[1]

On the fire scene, every firefighter has a perception of how the strategy should be formed and how the tactics should be applied to attain the goals in the strategy. These perceptions are often very different in the mind of one firefighter when compared to another— especially in stressful situations. An action is easily seen and measured while the effects of a safety item are often difficult to measure and the *prophecy of doom* may or may not be realized whether we change our operations or not. For example, we can see if a firefighter's Nomex hood is not donned properly—*Action*. In contrast, coming off a roof ventilation operation or pulling off a section of a wildland fire that may or may not cause harm if we stayed—*Perception*. The perceived risk as well as the actual risk is difficult to measure, not unlike trying to define how soft is soft or how hard is hard.

What can be done to further decrease the number of firefighter fatalities, injuries, and accidents? Some argue for increasing training opportunities. Others say obtain better equipment. Still others want to change the way we do our jobs. While all of these schools of thought would definitely make a difference, they are not the complete answer. The "most" correct answer lies in how we deal with each other and ourselves, the beginning concepts of this process you will learn within these pages.

Risk Homeostasis

Gerald Wilder, author of *Target Risk*,[2] explains why better equipment, more training, etc. will not altogether stop our firefighters from getting injured or killed. In his book, he outlines his Theory of Risk Homeostasis. This theory says, simply, that no matter how good you train firefighters or how good their equipment is, they will still get injured and killed. Why? All human beings on this planet have, ingrained in their minds, their level of acceptable risk. Everything we do as humans has risk; whether it is sitting in front of the TV eating chips and drinking beer (obesity, health problems, etc.), jumping out of airplanes (death, serious injury), or lying to the waiter about how old our kids are so they can *eat free*. We always, subconsciously, accept the level of risk we are comfortable with.

When on the highway, Driver A might drive 65 miles per hour, while Driver B drives 80 miles per hour, and Driver C drives 35 miles per hour. What is the main difference between these three drivers? You could say that the difference is in their age, gender, culture, experience, etc. Those factors are definitely a part of the bigger picture; but, the best reason for their actions is the level of risk they perceive in their actions and situations and how much they are willing to accept, i.e., their risk acceptance level.

Although who we are, our situation, and our environment play a big role in what we do, our level of risk acceptance dictates if and how we do it. You can say that Driver A is a middle-aged female in no particular rush to get anywhere; Driver B is a 21-year-old single male with a new, two-door red sports car going to see his girlfriend; and Driver C is an elderly female on the freeway at 5 PM coming home from bridge club. (But be careful when trying to stereotype people with regard to their risk acceptance levels—you could be wrong. I know a 70-year-old grandmother who drives like a 21-year-old.)

The problem with telling firefighters to, "Be Safe," is that they have a pretty high level of risk acceptance and they like it that way. Think about this: prior to the invention of SCBAs, firefighters did what? They went deep into a building to rescue occupants and to find the seat of the fire, right? *Wrong!* They did go into buildings but only for specific life safety priorities and minimal firefighting. Before fire resistant fabric and insulation materials in bunker gear, firefighters could stand only a couple hundred degrees of temperature for short periods of time. Their operations were definitely restricted by their lack of protective clothing and equipment.

Now, with state-of-the-art protective clothing, they regularly are exposed to higher temperatures and for extended periods of time. These two advancements (arguably some of the most significant since firefighting began), were designed to make firefighters safer and prevent injuries and deaths from fire and other hazards.

Imagine if we continued to fight fire like we did 60 years ago, with the equipment and knowledge we have now, our injuries and fatalities would be a fraction of what they were. But as humans, we used our new gadgets and training to their fullest and continued to go deeper into the fire to do our jobs "better and safer." Since we have these things to keep us safe for a longer periods of time in those dangerous atmospheres we can really be *firefighters.* It is not that we changed how much risk we will accept; it is more like we use our new equipment to even out the playing field so to speak. Unfortunately, we exposed ourselves to a whole new world of dangers. Our perception of the risks we are facing has changed due to the protective envelope we have placed ourselves in, but our risks have actually increased.

So, as you can see, we can put millions of dollars into new equipment, better training, and more research, but firefighters will use all of this against themselves, altering the way they do their jobs. They use their newfound knowledge and equipment, while placing themselves in even more dangerous situations while doing their jobs "better and safer." Do not misunderstand the argument here. We are not against developing better equipment, building our knowledge of fire and its effects, or increasing the training opportunities for our *brothers* and *sisters.* This will make our jobs easier, sometimes safer, and will serve our citizens well. We just believe that it will not stop the firefighter deaths and injuries that we see every year in the news.

The only way to really affect firefighter deaths and injuries is by modifying how we work together, especially under stressful conditions. That is what this book is about.

As firefighters, we face monumental risks in our jobs every day. There are a few ways of getting around these risks and therefore preventing many of the firefighter injuries and deaths—some acceptable, some not. First of all, we could drastically change the way in which we accomplish our jobs, in part, learning not to take risks above a defined level. In a sense, fire departments need to change their culture to embrace this philosophy. The way that a small rural department fights a structure fire should be quite different from the way an urban department does.

One task this book will not accomplish is to tell fire departments how to do their jobs. The fire service is too varied in how each department is organized, what resources are available to them, and the expectations each has, of themselves and their communities. It is up to each department to decide what best fits their local culture and use this information. We will tell you, however, that humans are humans are humans. It does not matter if you speak southern, northern, eastern, or western, when the ship hits the sand on an emergency incident, there is a pretty good probability that you will act in a predictable way—with minor variations. Do not take lightly what you will read within these pages.

Here is your first lesson in Human Factors/CRM—one we are sure you already know. *It is normal for people to summarily dismiss new information that they do not understand fully, that they disagree with, or that they think will never work in their particular area; i.e., any change in the status quo.* Keep an open mind. The material you will read is based on more than 20 years of research and experience in Human Factors by some of the best minds in psychology, and is peppered with more than 20 years of firefighting experience and research. This material will save lives. Put it to work for you, your fellow firefighters, and your department.

CREW RESOURCE MANAGEMENT

Think of CRM as an *"Owner's Manual for the Firefighter."* Due to some initial misunderstandings and misconceptions, the airline industry's first attempt at CRM was labeled, "Charm School." Pilots thought that they were being punished and were going to be taught how to be *warm* and *fuzzy.* That is definitely not the case. CRM is just about opposite of that perception. CRM is in place to give everybody involved the tools they need to get the information they possess to the people who need the information. Sometimes this is anything but glamorous and charming—sometimes it is flat out uncomfortable and almost painful.

CRM results in a sum that is greater than its parts—1+1=3, if you want to look at it that way. Each firefighter is only one person. But if all firefighters are taught, expected, and do work together, the sum of their individual parts can become much greater than just them together. For example, think of the incident commander. Even though he is in charge and has years of training and experience, he is still one person. One person can make a mistake, miss an important piece of information, or interpret the situation incorrectly. But using CRM Concepts, the synergy between his own skills and the skills of his firefighters can actually place him in all areas, observing all operations, watching out for everyone and making the best decisions possible. Hence, the leader has multiplied their effectiveness.

History of CRM in the fire service

On July 6, 1994, 14 firefighters lost their lives on Storm King Mountain near Glenwood Springs, Colorado. Much has been written about the fire. Perhaps the most in-depth analysis to date is in John MacLean's book, *Fire on the Mountain,* published two years after the tragedy. (Note: Ted Putnam, PhD, of the Missoula Technology and Development Center is currently working on his own complete report of the South Canyon Fire, *South Canyon: The Rest of the Story*).

Immediately after these deaths, the U.S. government appointed a group of high-level firefighting experts to examine the cause of the deaths. The conclusion of the official report is summarized on the next few pages.

Direct causes

The Investigation Team determined that the direct causes of the entrapment in the South Canyon fire are as follows:

Fire behavior

- **Fuels**—were extremely dry and susceptible to rapid and explosive spread. The potential for extreme fire behavior and reburn in Gambel oak was not recognized on the South Canyon fire.

- **Weather**—a cold front, with winds of up to 45 MPH, passed through the fire area on the afternoon of July 6.

- **Topography**—the steep topography, with slopes from 50 to 100%, magnified the fire behavior effects of fuel and weather.

- **Predicted behavior**—the fire behavior on July 6 could have been predicted on the basis of fuels, weather, and topography, but fire behavior information was not requested or provided. Therefore, critical information was not available for developing strategy and tactics.

- **Observed behavior**—a major blowup did occur on July 6 beginning at 4 PM Maximum rates of spread at 18 MPH and flames as high as 200'–300' made escape by firefighters extremely difficult.

Incident management

Strategy and tactics

- Escape routes and safety zones were inadequate for burning conditions that prevailed. The building of the west flank downhill fireline was hazardous. Most of the guidelines for reducing the hazards of downhill line construction in the *Fireline Handbook* (PMS 410-01) were not followed.

- Strategy and tactics were not adjusted to compensate for observed and potential extreme fire behavior. Tactics were not adjusted when Type I crews and air support did not arrive on time on July 5 and 6.

Safety briefing and major concerns

- Given the potential fire behavior, the escape route along the west flank of the fire was too long and too steep.

- Eight of the 10 Standard Firefighting Orders were compromised.

- Twelve of the 18 Watch-Out Situations were not recognized, or was proper action taken.

- The Prineville Interagency Hotshot crew (an out-of-state crew) was not briefed on local conditions, fuels, or fire weather forecasts before being sent to the South Canyon Fire.

Involved personnel profile

- The *can-do* attitude of supervisors and firefighters led to a compromising of Standard Firefighting Orders and a lack of recognition of Watch-Out Situations.

- Despite the fact that they recognized the extent of the danger, firefighters who had concerns about building the west flank fireline questioned the strategy and tactics but chose to continue with line construction.

Equipment

- Personal protective equipment performed within design limitations, but wind turbulence and intensity, and rapid advance of the fire exceeded those limitations or prevented effective deployment of fire shelters.

- Packs with fusees taken into a fire shelter compromised the occupant's safety. (Fusees are like magic sticks that you carry around as a wildland firefighter. When you need to light a backfire or blackline operation, you light them like a match and they burn for about 10 minutes.)

- Carrying tools and packs significantly slowed escape efforts.[3]

The rest of the story on South Canyon

The report, while generally correct, told only part of the story. Very professional, very competent and highly trained firefighters made some very dangerous decisions, which resulted in the death of 14 firefighters. Not only were these fatal decisions made by firefighters on the mountain, they were also made by Bureau of Land Management (BLM) managers in the administrative setting far removed from the actual incident. Dr. Putnam refused to sign the report issued by the accident investigation team because of the failure to address these organizational failures that had occurred. On February 15, 1995, Dr. Putnam released a paper, which was not specifically a report on the South Canyon Fire, but used it as a current example that emphasized what he had been analyzing for some 20 years—Human Factors. In the paper he made the following bold and insightful statement.

> *The fatal wildland fire entrapments of recent memory have a tragic common denominator: human error. The lesson is clear: studying the human side of fatal wildland fire accidents is overdue.*
>
> *Historically, wildland fire fatality investigations focus on external factors like fire behavior, fuels, weather, and equipment. Human and organization failures are seldom discussed. When individual firefighters and support personnel are singled out, it's often to fix blame in the same*

way we blame fire behavior or fuels. This is wrong headed and dangerous, because it ignores what I think is an underlying cause of firefighter deaths—the difficulty individuals have to consistently make good decisions under stress.

There's no question individuals must be held accountable for their performance. But the fire community must begin determining at psychological and social levels why failures occur. The goal should not be to fix blame. Rather, it should be to give people a better understanding of how stress, fear, and panic combine to erode rational thinking and counter this process. Over the years, we've made substantial progress in modeling and understanding the external factors in wildland fire suppression, and too little in improving thinking.[4]

Based on Dr. Putnam's recommendations and others, smoke jumpers began to implement CRM techniques. The nation's wildland fire agencies commissioned a three-phase study to assess the factors Dr. Putnam set forth in his paper as well as others.

The fatalities at Storm King Mountain affected the entire firefighting community. Command officers throughout the country studied the accident investigation team's report and then, Dr. Putnam's analysis. This author, inspired by Dr. Putnam's report, began to research and develop a CRM program for the fire service. As I began my research, I found the fire service was about 25 years behind the airline industry, which had been researching these issues for many years. I found pockets here and there where CRM training was creeping into the fire service, but no one had taken the issue by the horns and addressed it in a comprehensive fashion once and for all.

CRM in the aviation industry

In the late 1970s, an L-1011 crashed in the Florida Everglades. The plane crashed when the flight crew became preoccupied with changing a burned-out landing gear indicator lamp. While the crew members were all working on changing the indicator lamp, they failed to notice that the altitude hold function had been accidentally disengaged, and the plane simply flew into the ground, killing all on board.

In the same month, a B-737 crashed while attempting a go-around on an approach from Chicago's Midway Airport. The crew became preoccupied because the flight data recorder light became inoperative, and they lost track of where they were. On the initial approach, the crew deployed speed brakes because the plane was going too fast, was too high, and was not configured for landing. The pilot decided to go around. However, as a result of extreme time pressure, he forgot to deactivate the speed brakes and crashed the airplane. These two crashes served as a wake-up call to the airline industry.

For many years prior to these two incidents, the cause of crashes was equipment failure. But as the equipment became more and more reliable, it became apparent that the human animal was also a cause of accidents in the air. Under the leadership of Robert Helmreich from the University of Texas, Richard S. Jensen from Ohio State University, Al Diehl from the NTSB, NASA, FAA, air carriers, and others, uncounted hours of research and millions of dollars were spent in developing the CRM program. It optimized a crew's interactions in times of high stress and little information, where the lives of many people are at stake.

One of the most famous CRM accomplishments in the aviation community was United Flight 232 when Captain Al Haynes used all the resources available to successfully land a crippled airliner—a feat no one has yet to replicate in the simulations.

The CRM programs originally developed for the aviation industry have evolved so they are now used in hospital operating rooms, on battlefields, in corporate boardrooms, on college football fields, and now in firefighting.

Why does aviation research apply to firefighting?

The fire service finds itself at the same point as the Commercial Aviation Community was 25 years ago. When commercial air travel began, most of the accidents and incidents involved failures of equipment, inaccurate instruments, engine failure, and design flaws. As a result of a concentrated engineering effort, those problems were nearly eliminated, but tragic airline accidents still continued at an alarming rate. In a study of the accidents, the aviation industry discovered the accidents were associated with various failures in command, communications, and crew coordination.

Since that time, the aviation industry has spent millions of dollars studying CRM, human interaction, and efficiency in emergencies. The results have been amazing. As a direct and measurable result of CRM in the airline industry, commercial aircraft incidents caused by human error have been virtually eliminated, and through efficient management of crew resources, equipment failures have not had the devastating results they once had.

The thought processes and situations that face airline crews are fundamentally the same as those that face fire crews. Difficult life and death situations, which are not experienced in day-to-day living, and require immediate and confident action in a coordinated fashion by a crew, are the situations both firefighters and airline crews face in emergency situations. The airline industry has done a good job training its people to face such situations. Until recently, the Fire Service has not. Now, the Fire Service has an opportunity to take advantage of the millions of dollars of research, and not reinvent the wheel on this issue.

Our firefighting equipment is fast and reliable. Our SCBAs are state of the art. Our bunker gear protects us from temperatures in excess of 1000 degrees. But our firefighters are still getting hurt. A focus on the human factors affecting firefighting will prevent injuries and death.

What is the status of CRM in the fire service?

The fire service has now found these proven concepts knocking at its door. Equipment is becoming more and more reliable. Firefighting techniques and strategies are becoming scientifically honed, and new technologies for firefighter safety are being brought to the market daily. At the same time, firefighter fatalities and injuries on the emergency scene have plateaued. With all the new technology on the market, why is it, then, that firefighter fatalities have not significantly decreased?

Although we have developed a new and in-depth understanding of fire and our equipment has become more and more reliable, we have not focused on the most important machine on the fireground—the human machine. The lessons learned by the aviation industry are the future of firefighter training.

Presently, several initiatives are moving forward to adopt CRM principles for the Fire Service. The International Association of Fire Chiefs, under the capable leadership of Garry Briese, established a top-level team from the aviation industry, the International Association of Fire Fighters (IAFF), and command officers from all types of fire departments to study the application of CRM to the fire service.

After the deaths of 14 firefighters (smokejumpers, Hot-Shots, and Helitac) on Storm King Mountain, in Glenwood Springs, Colorado in 1994, a movement was begun to study the human factors affecting wildland firefighters (see *http://www.nifc.gov/scanyon/execsumm.html* for a description of the events). The third phase of the federally mandated Tri-Data study—Wildland Firefighter Safety Awareness Study—adopted 86 goals and more than 200 specific recommendations for improving the organizational culture, leadership, human factors, and external influences that affect wildland firefighter safety (*http://www.nifc.gov/safety_study/phaseIII.html*). All of the Federal wildland firefighting agencies were involved—Bureau of Land Management (BLM), United States Forest Service (USFS), Bureau of Indian Affairs (BIA), United States Fish and Wildlife (USFW), National Park Service (NPS), and the National Association of State Foresters (NASF).

Recently, the National Wildfire Coordinating Group developed an introductory training program on CRM principles for the line wildland firefighter. The Campbell County Fire Department in northern Wyoming was probably the first in the nation to adopt a comprehensive CRM program for its department. Finally, the International Association of Fire Chiefs (IAFC) has begun a major international initiative to adopt aviation CRM principles for the Fire Service. Departments in California, Georgia, Arkansas, Kansas, Wyoming, and Texas have begun CRM training. With the successful adoption of these programs, it will not be long until CRM principles filter into all phases of emergency services.

On the fireground, the mission should take into account three *legs* of the tripod for a successful, safe operation. The first is the *fire*. What is it doing, where it is, etc. are all things we think about now. The second is the *plan*. What are we doing, is it affecting the fire leg or is the fire affecting the plan, and others. The third leg is the *people*, CRM skills, interactions, and teamwork—how they are implementing the plan, etc. Without one of these legs, the tripod can fall; sometimes it is deadly (see Fig. 1–3).

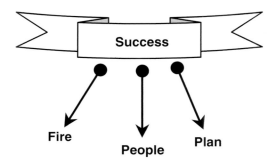

Fig. 1–3 The Tripod of Success

CRM concentrates on improving the *people* side of the tripod. By strengthening this leg, the *plan* leg is also improved and strengthened. So, even though the *fire* leg may collapse, the other two legs will be strong enough to prevent the tripod from falling on an unsuspecting firefighter.

Figure 1–4 shows the importance of using the skills and knowledge of all firefighters on a team, working towards a common goal, and supporting one another. If a firefighter and an officer are both having a good day, working together toward a common goal, the outcome is most often positive, or successful. If one or the other is a little *off* the outcome is unpredictable. If the operation is normal (standard) the outcome is usually positive; but, if something unexpected (non-standard) occurs that requires high levels of interaction and teamwork, the outcome could be negative, or unsuccessful—even deadly. Also, if both team members are having a bad day and not utilizing CRM concepts, the outcome is surely a disaster.

Officer Performance	Firefighter Performance	Outcome
+	+	+
+	–	Standard Incident + Non-Standard Incident –
–	+	Standard Incident + Non-Standard Incident –
–	–	= DISASTER

Fig. 1–4 Expected Team Outcomes from Individual Performance

THE CARDS WE ARE DEALT

In life, as in the game of poker, we have to play the cards we are dealt. Unfortunately, as a fire service, we are constantly putting our command officers in the position of hoping to draw to an inside straight. Fortunately for most of us, we get that lucky card a lot of the time. With the implementation of a proper CRM program, the need to draw to the inside straight can be minimized.

Human beings are animals. We all have weaknesses and strengths in perception, in knowledge, in preconceptions, and in thought processes. We have structured our fire command system around the assumption, that our fire commanders, because of their knowledge and experience, will not do the natural human thing, and make mistakes. The better practice would be to engineer into the system the assumption that our commanders are going to make mistakes, and to develop a system that traps those mistakes.

The Swiss Cheese Model of Errors

James Reason, the noted British psychologist, developed a model, which describes how CRM can be used as an error trapping system.[5] In his theory, a human being is like a slice of Swiss cheese. The human being has lots of capabilities, but there are holes in those capabilities. Where a hole exists, an error can pass through the Swiss cheese and result in an accident, injury, or fatality.

For a novice firefighter, the holes in the cheese are larger than someone who has lots of experience and knowledge. As we gain experience and learn more and more about firefighting, some of the holes get smaller while others—like preconceived notions and hazardous attitudes—may get larger. We work to train our command officers so that the holes get smaller and smaller and smaller (see Fig. 1–5).

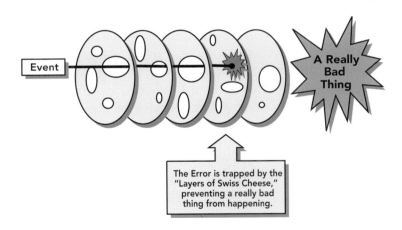

Fig. 1–5 Reason's Swiss Cheese Error Model

Unfortunately, there is nothing we can do to eliminate those holes in a person's ability to command a fire. What we can do is to engineer into the system, traps for the error. In other words, we need to build into the system checks and balances for the leader. By inserting additional layers of cheese into the system we can trap the errors. Those layers of cheese that we insert are the concepts taught in CRM, situational awareness, communications, leadership, followership, and decision-making skills. By inserting those layers, an error, which might pass through two or three layers of Swiss cheese, will be trapped by another layer before that *something bad* happens.

CRM encompasses so much information it is difficult to decide where to begin and agree where to end. Humans are complex animals, and years of study can just scratch the surface of understanding. Under stress though, humans can become quite predictable in their reactions. In this book we will look at some concepts to begin the foundation of understanding and open the floodgates for additional research and study if you so choose.

OVERVIEW OF CHAPTERS

In chapter 2, we look at what it takes for an organization to change its culture, embrace a CRM style of operation, and what effects it has in the short- and long-term. It is definitely a challenge to implement CRM, and change the attitudes of firefighters to embrace this new, open method of operation, but it is well worth it.

In chapter 3 we teach you how to establish a mission objective instead of doling out tasks constantly. You learn to allow subordinates to use their own knowledge, skills, and perception of the situation to make better and faster decisions that result in more efficient and safer operations.

In chapter 4 we discuss situational awareness. Did you know that your perception of reality is not correct, especially in the situations firefighters are faced with on an emergency incident? We tell you how to make your perception of the situation more closely match the reality, how you lose your awareness of the situation, and how to regain it and keep it. Scotoma, the inability to receive information that does not seem to be important or highly dangerous, is a major problem with firefighters on emergency scenes; it can limit your perceptions and affect your actions.

In chapter 5 we discuss communications. In almost every deadly accident or injury there are links that implicate a breakdown in communications. We discuss what to expect of you and your team members during times of stress and how to fix a majority of the communication problems within your department and operations.

In chapter 6 we discuss leadership. This is a subject that seems overdone in trade publications and theories. We actually build on those skills you already possess and teach you how to use your knowledge of fire, organization, management, and people within the context of CRM to improve the safety and success of your operations, as well as the morale of your people.

In chapter 7, we introduce *followership*, an entirely new concept to most paramilitary organizations. No more is your job as a subordinate to keep your mouth shut and follow orders. You actually learn how to be the foundation that your department builds its successes on and how the individual decisions you make every day in the house and on the incident affect the operation and your safety. For example, did you know—

> *Of the 40 heart attack victims in 2000, 11 had previous heart problems, usually heart attacks or bypass surgery, and medical documentation showed that four had severe arteriosclerotic heart disease, one was hypertensive, and one was diabetic.*

> *During the past 24 years, medical documentation has been available for 642 of the 1256 heart attack victims. Of those, 49.4% had had previous heart attacks or bypass surgery, and another 30.5% had severe arteriosclerotic heart disease. Another 12.8% had hypertension or diabetes.*[6]

In chapter 8 decision-making under stress is addressed. This teaches you how your memory, information processing, and decision-making skills are affected by stress. Knowing what to watch for and how to prepare yourself before it happens are vital to your success and your crew's safety.

Chapter 9 helps you and your department keep CRM alive without multitudes of annual refresher trainings and such. An easy add-on to what you probably already have in place will ensure that your program stays strong and your people stay safe.

Finally in chapter 10, we discuss some strategies for implementing CRM in your organization and what to expect.

We truly believe that this material will save firefighters' lives *and* improve our operations nationally. Keep an open mind and read this book with your specific department and operation in mind. This is not a one-size-fits-all book—just some ideas for you to take, adjust, and apply to your situation. Use it and we are sure that you will make a difference—it already has in many other areas.

REFERENCES

1. Report: *Fire in the United States*, 1989–1998, Twelfth Edition, produced by Tri-Data Corporation under contract from the United States Fire Administration.

2. Gerald J. S. Wilde, PhD, *Target Risk 2: A new psychology of safety and health. What works? What doesn't? And why....* PDE Publications. 310-5334 Yonge Street. Toronto, Ontario, Canada M2N 6M2. (2001) Author Address: Professor Emeritus of Psychology. Queen's University: Kingston, Ontario, Canada K7L 3N6. *wildeg@psyc.queensu.ca* : 6.

3. Report, *South Canyon Fire Investigation of the 14 fatalities that occurred on July 6, 1994 near Glenwood Springs, Colorado*, NWCG, Boise, August 17, 1994.

4. Putnam, T., *The Collapse of Decision Making and Organizational Structure on Storm King Mountain*, USDA Forest Service: Missoula, 1995.

5. Reason, J., *Human Error*, Cambridge University Press: Cambridge, 1990.

6. *2000 U.S. Firefighter Fatalities*, Fahy, Rita, PhD, and Paul R. Leblanc, *NFPA Journal*, July/August 2001, 70. Reprinted with permission from *NFPA Journal* (Vol. 95, No. 4), National Fire Protection Association: Quincy, MA, 2001.

ORGANIZATIONAL SAFETY CULTURE

2

Any man worth his salt will stick up for what he believes is right, but it takes a slightly better man to acknowledge instantly, and without reservation, he is in error.

—GENERAL PEYTON C. MARCH

EVERY FIREFIGHTER is responsible for personal safety, and the safety of every other firefighter. Given that responsibility, every firefighter must have the authority to protect the safety of every other firefighter. When a balance between the responsibility for safety and the authority to protect safety is reached, the firefighting organization reaches a true culture of safety. Ignorance of or perhaps simply denying the traits of human beings working in stressful organizations is the barrier between the true culture of safety and a safety program that may be designed to fail.

The distinction between a safety program and a safety culture is important. A safety program is another in an endless series of initiatives designed to make the individual firefighter aware of the risks inherent in firefighting. On the other hand, a safety culture grows from the very roots of the organization. Instead of a superficial topical group of ideas, the safety culture begins with the mission analysis of the department and extends into every structure of the organization. Safety is the first thought, and all other decisions, operations, purchases, and programs revolve around the culture of safety.

Phase III of the Tri-Data study[1] noted the needs for a safety culture in the wildland firefighting environment. After conducting a comprehensive study of the wildland arena, from the top down to the line firefighter, the authors of the study found a pervasive cultural disregard for safety concerns.

> One-fifth of the responding employees felt that their supervisors do not listen when they voice safety concerns. Many people also expressed concern about the potential for reprisal if they speak out about safety problems. Additionally, evidence suggests that firefighters do not generally regard speaking out about safety issues as a personal, individual obligation. This led to Goal 2.

> Goal 2. A Code of Conduct should be established in which employees should have both the right and the obligation to report safety problems, and to contribute ideas on their safety to their supervisors. The supervisors are expected to give the concerns and ideas serious consideration.

The authors could find no such comprehensive study of the structural firefighting service, but our experience indicates there is significant crossover between the structural and wildland communities.

The true measure of a professional fire department is its regard for safety. No matter what the fire department does, the fire eventually goes out. No matter how clean the trucks are, the sun rises the next day. As organizations, what we members can change is who goes home healthy and who does not go home at all. We can change the way we serve and protect ourselves and others. In a world of extreme and unknown dangers, the only true professional fire department is the one that looks out for the long-term health and safety of its members and the public it serves.

ORGANIZATIONAL CULTURE CHANGE— NOT MORE PROGRAMS

The fire service is a paramilitary organization

The implications of that statement are mind-boggling. *Paramilitary.* What is our mission? What is our commitment to our mission? Imagine yourself in a military combat situation. The general gives an order to take Hill #5834, and the order trickles down the chain of command to you—the one with the rifle *and* the risk. What planning has gone into this order that makes you run up the hill? What feedback is expected from you that travels up to the level of general?

Think about it in a wartime situation, because that's what we do on a fire scene, basically wage war on a fire. Many of you know that when the general makes a strategic decision, there are serious implications—possibly your death! Even successful operations have dire consequences. A military operation may be of such strategic value that a 10–15% soldier mortality rate is acceptable.

What is the acceptable mortality of your engine company? How many of your crew members should be sacrificed for the mission? You have a lot at stake personally in this operation. Could you call back to the general and say something like, "Uh, General, this is really tough going, can we go around the other side of this hill?"

The Civil War battles are an example of what dedicated military men will do for a cause they hold dear. What horrors must have been met on the battlefields of the 19th century when entire lines of men rushed madly toward their objective, and literally thousands fell, one after another, stepping over their fallen comrades to reach their objective. These were heroic, dedicated individuals following orders. To what degree are you willing to lay down your life for something you believe?

We are not a paramilitary organization in many senses of the word. Our acceptable losses are far less. Our mission analysis comes from a different perspective. We do the best we can with what we have, knowing that we did not create the situation, and anything we do has a chance of making things better.

The sad facts show, however, our current emphasis on *safety programs* hasn't made a huge difference in decreasing the number of firefighter fatalities and injuries that we endure every year. The increased amount of personal protective equipment hasn't significantly decreased those numbers either or for a significant period of time. The difference is going to be made at an organizational and individual level—period! We will make the difference by understanding why we do stupid things that get firefighters or ourselves hurt or killed. We will make the difference by knowing what to expect when it's coming down around our ears. By using *human factors training*, we will make a difference.

The organizational change

For the fireground to become a safer place, the organization itself needs a major shift. This has been proven in numerous incidents and studies inside and outside the fire service. Organizations in America have found that there is a definite chain of events that leads to *incidents* (small, no injury, little cost problems), *accidents* (some injuries, more cost), and *fatalities*. That chain of events, the error chain, once was wrapped around employees' necks by management. Now, researchers and investigators have found that the chain of events travels from the actual employee involved up to the highest levels of the organization. The 14 wildland firefighter fatalities on Storm King Mountain in Colorado in 1994 show a lengthy chain of events and errors extending to the highest levels of government—all the way to Washington, D.C.

The change in an organization cannot be just lip service. There must be a concentrated, continuous effort made to change the organizational culture of the workplace. As proven with the *safety program mentality*, saying we want safe workplaces and actually having them are two very different phenomena. The current safety program mentality attempts to convince employees to *be safe*, but may not give them the tools they need to report unsafe activities and behaviors. The safety program mentality often costs a lot of money to implement with new training programs and better and safer equipment. Also, the safety program mentality usually does not address the needs of an emergency service department that immerses itself in risk every time the department performs.

The solution to throw money at the fire isn't a very long-lasting solution and quickly degenerates into a Black Hole where your entire budget will be consumed—constantly buying better equipment and better safety programs. When you discuss safety, of course, the equipment quality, type, reliability, quantity, etc., are very important.

But a safe working environment doesn't begin with thousands of dollars of highly engineered equipment. It begins with the person who will use that equipment. Firefighters can be trained to fight fire *safely* with a bucket and a mop or they can be trained to fight fire *safely* with a $1 million dollar aerial apparatus and a $50,000 personal protective ensemble.

It comes down to one question, "How effective can we be with what we have?" Unfortunately, in this day of increasing demands for service and fire managers who are attempting to build a politically secure position, the question that is usually asked is, "What can we do to improve our services?" The differences, although literally slight, can have major implications in emergency situations.

Take for instance a typical structure fire. Put that room and contents fire in the context of a rural home 20 miles from the nearest fire department. Then, put that same room and contents fire inside the five-minute response time of a highly organized fire department. In the rural setting, a firefighter with limited water supply, limited or no protective bunker gear, and no SCBA can still perform a service and do it safely.

In the metropolitan area, the firefighters are generally better equipped *safety-wise* to deal with these types of fires and provide the service. Keep in mind the main goal of the operation: *Everyone goes home safe.* Who accomplishes the goal? Both fire departments can effectively achieve their goals. Now, it would be nice to prevent total loss of the contents of the home and exposures, but that is *not* our *main goal!* Think of it this way.

> *The rural fire department does not put out the fire before it destroys a family's home and the garage exposure. The firefighters stayed outside and attempted to put out the fire from the exterior because they don't have any bunker gear. No one was injured during the fire.*

Or

> *The metropolitan fire department stops the fire while it is still in the original room. Smoke damage is light inside the home and the family will be back within a week. But one firefighter was severely injured when he slipped with the 35' ladder he was carrying by himself. Because of reduced staffing and the actual or perceived speed at which the operation must continue, this firefighter did not wait for help and therefore will suffer through a partial disability the rest of his life, and will not be able to fight fire any more.*

Within the context of these two incidents:

- Who achieved their *main goal?*

- Who provided the *best* service to their citizens?

- Who provided the *safest* service to their citizens?

Be very careful when you answer these questions. Remember the military example? Do we sacrifice firefighters to provide for strategic objectives? Is a firefighter's health worth the price of a saved home?

What you will learn in this book about the implementation of the concepts costs basically nothing. That's right! Basically nothing. Beyond a few hours of training, you can implement this entire book into an existing department basically *for free!* Also, if there's a more compelling reason to adopt this concept of a safety culture, try this on for size—You'll get a *free set of Ginsu steak knives!* Not really...*but it can cut your death and injury rates among employees to a mere fraction of what it currently is!* (No pun intended).

The commercial aviation industry saw a 70–80% drop in accidents and incidents in a very short time after implementing human factors training.

The catch? The catch is that the implementation of CRM principles must be a calculated and coordinated effort. CRM is not a magic bullet. It will not affect all aspects of your organization immediately; and even after full acceptance into a department, there will be times when CRM does not work. CRM *only* works effectively in functional organizations with technically proficient personnel. Functional organizations must have continuously supported technical proficiency, individual responsibility, accountability, and discipline. Without these rungs in the ladder, CRM is doomed to drag your department lower—and lower than it has ever been.

THE ERROR-FREE WORLD

Like the lottery, you know it exists and you may even know someone who has won, but the existence to you is probably as foreign as a summer home on the French Riviera. Even people who have won the lottery and now have millions of dollars don't enjoy the place where they are. Likewise, the fire service wouldn't attract the type of people needed if there weren't any chance of risk or heroism—they just wouldn't like it.

In the fire service, we can never become *risk-free. In* the fire service, we can never become independent of human behavior. So, we will never become *error-free, injury-free,* or *fatality-free.* What we need to concentrate on is developing a system to reduce the errors that we encounter in the fire service. Incidents, accidents, and fatalities occur when multiple individuals, often in multiple levels of an organization, make one or more errors, nobody stops the progression of these errors, and then something bad happens. By using CRM principles, we are providing barriers to the proliferation of errors. The more people we have watching for errors (barriers), the fewer unbroken error chains exist and the fewer and less severe are the problems that occur.

Probably as much as any profession in the world, firefighters enjoy their jobs because of the level of risk. CRM doesn't preach an error-free culture. It does tout fewer stupid errors missed because someone became overloaded, made a bad decision, wouldn't accept important information, or one of the hundreds of other problems CRM will help reduce. In fact, CRM may enable you to take more risks and perform more dangerous operations because you can count on the input and monitoring of all the players, from the chief to a probationary firefighter.

THE INDIVIDUAL RESPONSIBILITY

After the organization has bought into the principles of CRM, the individuals must. Without the support of the individuals, CRM is doomed. How many times have you seen the highest levels of an organization put forward policy or procedure as law, only to have it fail because of the lack of support from the individuals? Individuals hold the key to success of CRM in the organization. Support them and adjust the principles to match your organization and operations and CRM will be successful.

The principles that will be explained inside the covers of this book are guidelines. Much research and learning is taking place right now in the *human factors community*. Many smart people are addressing how people interact with one another and with machines—both with and without the stress of emergency situations. All these disciplines will become very important to you as individuals. Don't hesitate to continue your human factors training outside the fire service scope. Astronauts, commercial aviation pilots, nuclear power plant operators, military organizations, surgical teams, Olympic athletes, college and professional athletic organizations, scientific expeditions, adventurers, and many other areas are using these principles to tune their teams to the highest proficiency possible. Use their experiences and knowledge to further your personal and organization's knowledge of the way human brains work.

BARRIERS TO IMPLEMENTATION

Power structures, organizational inertia, budgetary constraints, and fear of change all act as barriers to successful implementation of a safety culture.

While we know many departments have a true organizational safety culture, our focus is on those who don't and why. We understand that there may be 10,000 perfectly safe fireground operations; our focus has to be on the one operation that is not safe.

The biggest barrier to implementation of a culture of safety is the power structure in the organization. Traditionally, firefighting has been modeled after paramilitary organizations. Command officers gave commands, and line firefighters did as they were told. The fire scene worked like the old Alaskan adage, "The lead dog is the only one with the good view." Firefighters trusted in the decisions of the lead dogs, and the lead dogs took the safety of their followers very seriously. The system works only as well as the lead dog can lead, which, most of the time, is very well.

With position came power. Situations began to arise in which the lead dog could wield authority without being questioned. As long as the lead dog knows everything that the following dogs know and sees everything that the following dogs see, it is the best system available. The lead dog is responsible for everything, and if something goes wrong, we simply get a new lead dog. The system is simple and has lots of chances for advancement because it is constantly replacing lead dogs as things go wrong.

Developing a culture of safety, however, encourages other dogs on the team to question policies and practices. The following dogs are trained to a level of technical proficiency and then taught to think on their own, to advocate, and to challenge. The comfortable and unquestioned seat of the lead dog changes. Instead of having all the power and responsibility for decisions, the lead dog's authority is diluted. In some ways, the job becomes easier, because there are more active minds working on the problem. In some ways, the job becomes more difficult, because the lead dog no longer has absolute authority and has to be able, if time allows, to accept input from others and modify the approach. While the ultimate authority of the lead dog is still there, responsibilities for outcomes belong to the entire team.

The dissipation of pure authority from the leader of an organization tends to be the biggest impediment to the adoption of a safety culture in departments without such a culture.

The second factor that prevents the adoption of a safety culture in an organization, is operational inertia. Like the power structure, which is jealously guarded, organizational inertia proceeds forward without reason. People do certain things, certain ways, time after time. Pretty soon, the reason for doing the things is lost, but the tradition is still there.

The best example of such a practice is the woman who made ham for Christmas dinner. Before she put the ham in the oven, she would cut the ends off of the ham. One year, after watching this practice every Christmas of their marriage, the husband asked, "Why do you cut the ends off the ham?" The woman answered, "I don't know. My mother always did it this way." A few weeks later, the woman was talking with her mother. She asked, "Mother why do you cut the ends off the ham before you bake it?" The mother answered, "My pan is too small. The ham won't fit unless I cut the ends off."

Continuing a mindless task without asking why is like throwing perfectly good ends of the ham into the garbage. To change the routine takes a certain level of energy. Adopting a new program without a track record is difficult in the face of the grizzled old veteran, standing beside the coffee pot, saying, "We done it this way for 40 years, and no one is the worse for it. We've tried all kinds of crap and it has never worked. Why should we try something else now that will never work?" Only a dedicated management that has taken the time to market the culture to the organization will succeed in changing the attitudes of the culture.

The next barrier to successful implementation of a safety culture is the budget. The city fathers just do not have enough money in the city budget to purchase reasonable protective equipment and still build a monument to themselves. Of course, they have enough money in the budget to pay workers compensation premiums for injured firefighters and to cobble together old equipment to get the job done. (Our cynicism is justly earned. Try to tell the city fathers that it is important to send a firefighter to an aviation psychology conference and see the *happy* faces of the community leaders.) Talking about a nebulous concept of firefighter safety to a board of budget-conscious elected officials is a difficult task, particularly when it is not specifically quantifiable. Without gains or losses the accountants can tally on the profit and loss statement, a safety culture is difficult to sell.

However, if the experience of other industries is any measure, the airline industry's 80% reduction in airline incidents, along with injuries, lawsuits, dead bodies, bad publicity, and decrease in business has justified a look from the airline industry management. Also, mandated is a federal requirement from the aviation administration that CRM become the basis for the safety culture. Safety culture is the predominant culture in the aviation industry. Those who disregard safety for the bottom line are termed *rogue operators,* and are the subject of much disdain in the aviation industry.

A safety culture in the fire service these days can be quite the opposite. We pride ourselves as the folks who are the ones running into the burning building while everyone else is running out of the building. We train our people to look death right in the face and laugh. Most of the time, we get the last laugh.

Unfortunately, more than 100 times a year, death gets the last laugh. If the aviation industry's experience is any measure, we can reduce those deaths by 80%. There would be 80 fewer fire-fighter funerals, 80 fewer grieving families, and 80 fewer lost colleagues, and eighty fewer one-parent families each year. This can happen if only we can change the culture of the fire service from rugged danger cowboys into thinking, communicating, evaluating and monitoring, professional firefighters.

The culture change begins at both ends of the spectrum. National leaders in the fire service need to spread the message. At the same time, the cultural change needs to be inculcated at the entry firefighter level. When the two messages meet in the middle, we will have truly created a culture of safety.

CREATING THE SAFETY CULTURE

Five steps are important in creating a culture of safety. First, the organization must build trust among its members. The honest sharing of safety information must be encouraged among all members of the organization. Junior members of the organization must not fear reprisal or retaliation from superiors. Senior members of the organization must not be punished for pointing out a problem where one was covered up successfully in the past. Only in an atmosphere of honest sharing of safety information and honest steps toward resolution of safety concerns can a culture of safety be developed.

Second, the organization must adopt a non-punitive policy toward error. Error is part of the human condition. Human beings make errors. For years we have tried to deny the simple fact of life, *people make mistakes*. We try to train away errors. We say to our leaders, "You are now a leader, you can no longer be human. You can make no mistakes, even though they are wired into your being."

The fact of the matter is—our leaders are human beings. They do make mistakes. Instead of denying the obvious, we should plan on people making errors, and program checks and balances into our system to catch those errors. The only way to accomplish such a balance is to know when and how those errors occur. To gain that knowledge, we must ensure that each error is reported. If we punish human beings for reporting errors, we deny ourselves the information necessary to program checks into the system. We know humans will make errors. If we deny ourselves the information to stop an error, the error will occur over and over and over again; and one fateful day, it will combine with a series of other errors to kill one of our friends.

An organization should develop a non-punitive policy toward error, which does not punish those who are trying to accomplish their jobs in accordance with the regulations and the SOPs. Then the organization has the opportunity to address the errors and to develop a safer organization.

Third, the organization must demonstrate a willingness to reduce error in the system. All of the emphasis on safety and development of error reporting does not mean anything if the organization simply gives lip service to trapping the error. Honest, concrete steps must be taken toward the reduction of error, within a reasonable time after reporting the error. When members of the organization see positive evaluation, changes, and steps toward solving the problems in the organization, the members will begin to address the errors themselves.

Fourth, the organization must provide training in error avoidance, detection, and management strategies for the crews. In order to successfully combat error on the fire scene, the firefighter must be given the appropriate tools and language to recognize and communicate the problem. If the crews are given the tools to address the issue, the authority to address the issue, and the knowledge of what the issues are, the issues will be addressed. No firefighter goes out for a day on the job and says, "I'm going to get myself or someone else hurt today." The natural drive toward self-protection will motivate the crew into safety action. Given the tools, firefighters will protect their own.

Finally, the organization should provide training in evaluation and reinforcing error avoidance, detection, and management. A clear, honest debriefing of incidents, showing both the good and bad points, serves as a perfect tool for meeting this goal. The NASA Space Flight Resource Management Program contains an excellent debriefing tool and checklist for evaluation of both training and missions. Astronauts and trainers alike are instructed on the debriefing process and all share a common goal of safety. The culture is very safe in an organization where a small mistake can mean certain death of both the astronaut and the organization.

The firefighter faces the same challenge as the astronaut. Our deaths, though not as spectacular, are just as real.

REFERENCES

[1.] *Wildland Fires Safety Awareness Study: Phase III*, "Implementing Cultural Changes for Safety," Tri-Data Corporation, 1999.

3

MISSION ANALYSIS AND PLANNING

IF YOU OPERATE UNDER THE PREMISE that the more you know the better off you are on an emergency scene, you can understand how important mission analysis and planning is. Mission analysis and planning are the processes that take place using available information, missing information, resources, prior experience, developing strategies and tactics, assigning duties, etc., before, during, and even after an incident. Mission analysis and planning occur to some degree at every level, on every incident, and in every department. The trick is to perform them in a thorough manner that lends itself to an effective, safe operation and opens the door for effective CRM.

You should arguably take the mission analysis function to all levels of the organization, to external organizations (such as funding agencies, policy makers, inspection agencies, etc.), and to all different times— past, present, and future. What we will concentrate on in this chapter is the actual incident. We all know the responsibilities of preplanning, zoning, codes, etc. While important, these are beyond the realm of the operational CRM we will discuss. I believe we all realize that these types of activities establish a foundation for our on-scene activities, safety, and results.

While constant training and scenarios are very important to a properly functioning department, there is no way we can foretell how future incidents will play out. In fact, some amount of danger exists in preplanning and *burning down buildings* in the classroom if too much (or the wrong) emphasis is placed upon them.

When students are drilled about the exact location of a fire that will overtake the Smythe Building on Longmont *sooner or later,* they begin to become rigid in their thinking and become inflexible to the reality of what they may find at the actual scene. In fact, these problems may even transfer to other incidents if we are not careful. We all probably have buildings in our jurisdictions that are going to burn and we probably have a good idea why. When we train, we discuss where the fire probably is going to start. We discuss the death-trap, bowstring trusses that will collapse early with no warning. Our training includes the problems of wall coverings, roofing materials, *anti-gravity mechanisms,* and all those other myriad technical factors.

The problem with that type of rigid training is that if and when the building ignites, we have pre-destined our operations *and* our effectiveness in many cases due to preconceived notions that may or may not match reality. For example, a fire in a targeted building may not have started where we had anticipated but rather in a garbage can. All of the training had been directed toward a fast-moving fire that would require an aggressive exterior operation due to the construction type. Under stress, the firefighters may resort to their learned behaviors and knowledge and overlook the facts (the small fire) and begin to establish a defensive attack.

There are only so many factors that we can know for sure—both in advance and during the incident. We cannot plan away the unknowns that exist in our jobs and our scenes. We must not be bent on a single vision of the problem that most certainly leads to a single solution for what could be the wrong problem. It is like getting the right answer but using the wrong formula. Sure, you may have gotten it right, but will it happen again? Will the formula work again or not? We need to make sure we are using the right formula for our jobs, and preplanning scenarios, coffee pot chatter, and the like are a very small piece of our on-scene formula. We must use the entire formula to come up with a correct answer—one that is safe and effective.

THE FORMULA

The formula for successful fireground operations contains many elements. Each element, combined with the others, will determine the results of a fireground operation. The formula is the operation. The result is the answer. So, our answer is going to be, "Everyone Goes Home Safe." We have pulled safety out of an operational setting and placed it in a planning setting where it should be. We are not trying to be safe *while* performing a tactic; we are planning our tactics *to be* safe. This is a very important distinction and one we will discuss more in a later chapter.

We begin our mission analysis and planning by gathering our previously known information. Good leaders know what information is useful and which is inappropriate for this incident. They will use this information, not so much as a base, but as one rung on their ladder to an effective operation. If you can picture that ladder of success, one beam is the judgment of the leader and the other is the experience of the team. Those two beams constitute main supporting elements. The other information forms the rungs between the leader and the team. But both beams must agree (or at least be comfortable) with the rung being added before it is put in step (see Fig. 3–1).

If the leader places too much emphasis on a pre-remodel preplan of the building, the other beam, the team, can use their communication skills and say, "Wait a minute. We don't think that really applies anymore because the situation has changed too much. Let's not depend on that to tell us the room layout." If both beams agree, the rung is added and the organization is one step closer to the successful outcome. (Proper communication techniques of inquiry, advocacy, and monitoring are discussed in depth in the communications chapter.)

Other elements in the formula include experience, judgment, training, team trust and strength, resources, and size-up quality.

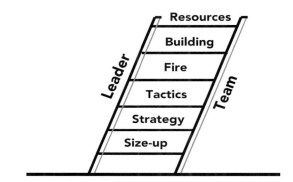

Fig. 3–1 Information: Rungs on a Ladder

An important change in the formula rules as they relate to the inputs is that any individual can be a resource with the correct information to input into the formula. The input could come from a firefighter, a citizen, a dispatcher, or any other resource. In the past, we have been too caught up in who has control of the formula and the inputs. Today, we should look at all sources of information as to the firefighters' abilities to apply the correct information to come up with the correct answer. This means leaders must be open to communications from all different sources and be able to apply judgment to the information instead of rank. Not all information is good, but it should all be judged on its merit and not just its source.

The greatest lesson in life is to know that even fools are right sometimes.

–Sir Winston Churchill

For the team as a whole to be effective, the team as a whole must be knowledgeable in what their duties are for their own task, to the strategy, to the operation as a whole, and to other teams. This means that people—teams—must be briefed before the operation begins; they must be kept informed on the status of the incident and the plan; they must be debriefed or critiqued after the operation; and then they must use this new information and knowledge in training and application.

The briefing

NASA tested 7500 flight crews and determined that the two most significant factors in the promotion of good CRM principles were

- The Captain giving a thorough briefing.

- The First Officer (subordinates) making inquiries and advocating his/her position.[1]

In the fire service we can do the same things.

- Always give a crew briefing and make sure instructions are understood.

- Always monitor, inquire, and advocate your position no matter what your level in the organization.

How many times have you been on an incident, performed your task, and gotten back to the station only to find out there was a major hazard/problem in another area that could have adversely affected you and your team? Ever been inside one portion of a building when another portion had a partial collapse? Ever been on a traffic accident when a drunk driver ran into an emergency vehicle parked behind your engine? Ever treated someone at home for a medical problem and then found out later that person had a contagious disease? Ever been on one division of a wildland fire and found out later at camp that a backfiring operation had been started next to your division?

The issue is, "You don't know what you don't know; and what you don't know could kill you." How can firefighters be sufficiently safe and effective if they do not know what they and others are doing, what hazards they are facing, what problems they must overcome, and what other support and resources they can expect, request or offer.

When we say that everyone needs a briefing prior to performing operations, we mean many different things. A chief officer will get a little bit different briefing than the engine crew. The briefings do not have to be very long and although they should follow some sensible flow, they do not have to be multimedia showcases. Simple, to the point, covering all the pertinent bases with time built in for feedback and questions will be enough.

Generally, the brief should include

- The desired results of the operation

- Expectations of individuals, teams, etc.

- Responsibilities of individuals, teams, etc.

- Available resources

- Micro-training opportunities

- Encourage input from team members

- Positive, proactive attitude

- Accountability[2]

Desired results

The wildland community is very good at defining desired results. These *goals* are written down and every attempt is made to support them.

Examples could include

- Ensure the health and safety of firefighters and the public

- Make efforts to utilize tactics and equipment that reduce environmental damage

- Keep the fire to the east side of Rocky Point

They may sound political or even redundant sometimes, but they serve a very important purpose—they allow everyone involved to see what the desired results of all their hard work should be. Without addressing the desired results, the operational plan sometimes gets off course, and we achieve something we did not set out to achieve.

Do not take for granted that your team knows the desired results. Their operational picture and your strategic picture are not always the same. Do not assume that just because we all showed up on big red fire trucks that our desired result is always putting out the fire. Make the desired results explicitly known to your team and any other involved parties.

Expectations

Expectations of individuals and teams on emergency scenes may sound a little redundant. We already know the desired results, right? Why do we need to go over what is expected of us, too? Like your mom always told you—*because!* There are just too many factors and variables to leave important items to the *Assumption Express*. We do not expect the IC to systematically go down the line and tell all firefighters individually what their specific expectations are for this particular call; that would just be crazy. "Let's see, Jones, wear your gear, don't trip over any curbs, and carry your tools into the house so you won't have to come back for them. Barrington, wear your gear and make sure you hold onto that hose line really well. Rastivson, don't forget to drink more water on this fire, okay?"

A multitude of on-scene expectations exist that are transferred to firefighters during trainings, meetings, etc. Above and beyond their duties as firefighters, they need to know what operational expectations you have for them that may be in question or above and beyond the norm.

What we need to get in the habit of doing, as a process of a briefing, is to let everyone know what these *extra* expectations or operational expectations are. A good example of this occurs when we operate under lightweight trusses. An expectation might go something like this, "I'll have dispatch mark time in 5-minute increments and notify me via radio. At those callouts, Engine 21, I expect a report on the conditions inside the building and your progress. If I don't stop the time, dispatch will set off an emergency tone at 15 minutes. That will be our cue to exit the building. Engine 21, I'll expect an accountability report when you reach the exterior."

This specifically sets the stage for what is expected of the people on scene. Other incidents will not be so structured. The expectation may be to handle the situation and to let the IC know if you need anything. Whether strict or liberal, formal or informal, expectations should be explicit. Do not wait around outside the building and complain about the lack of information or action. Be proactive in what you will need from the team and they can do the same for what they need from you.

Responsibilities

Occasionally, there will be times when the exact responsibilities of teams and individuals will not be apparent to everyone involved. Better to address those outright before running into problems later on when emergency conditions exist that make planning and communications more difficult. Make sure Division A knows that he is the entry point and that Division C is the secondary emergency exit. Make sure the Water Officer has the authority to order private tankers if necessary. Ensure the Medical Sector has the responsibility for rehabilitation and for on-scene response.

Most of the time, these priorities can be covered in training. But, if something is non-standard or you require something extra of the firefighters (especially if it is pivotal to the operation), make that known up front. If you want someone to take the responsibility for a certain operation, make sure you tell them that up front and also tell them what kind of feedback you need. Could it be that you just want a firefighter to handle it, or that you would like to approve it? Or should the firefighters just let you know when they do it? It varies from incident to incident, so make their responsibilities known. This helps to lighten the stress and workload on leaders and it reinforces good CRM skills and expectations.

Available resources

The fire service is very good at utilizing resources and acquiring what they need—mostly through legitimate channels—but sometimes that is not the case. On one particular windy day, we were in another jurisdiction for mutual aid on a wildland fire. When there was no apparent organization or plan, our team took control over what we could and communicated our needs to the agency's contact—the mobile IC. What we found was that we had all the responsibility and fire in the world, but no resources. Make sure that you and your people know what is available. It could make the difference between a 3-minute or a 15-minute interior attack or even a defensive stand. Make sure the teams know what is available and how long it will take to get there; it will definitely change their tactical plans and it will help to lighten the stress on the firefighters.

It is very stressful to arrive on-scene, and expect to perform a standard operation, only to discover that the usually available resources are not there. You can literally see and feel the firefighters' frustrations. This could deteriorate into an unsafe, supposedly heroic act that could have been avoided if only the briefing included the lack of available resources. Without any hard data to substantiate this, but with many years of experience, I would say that many injuries, fatalities, and unsafe acts are due to standard strategies and tactics being applied to non-standard incidents or ones without the proper resource support. Firefighters try to get the job done using skills and techniques that were forged for use with X number of resources and support, but when your resources are X divided by three, you have to know that ahead of time, plan for the shortages, and adjust your strategies, tactics, and attitudes.

MICRO-TRAINING OPPORTUNITIES

Experience is a hard teacher because she gives the test first, the lesson afterwards.

—VERNON SANDERS LAW

We learn from everything we do. Sometimes it is something we already know that we reinforce or sometimes we learn something new. Sometimes, we learn the wrong things and sometimes the right ones. But everything we do on an incident adds to our knowledge and experience base. This is true for almost every animal on the planet.

Take for example grizzly bears in and around Yellowstone National Park. After a few accidental run-ins with humans, some particular bears discovered through experience, that when weary, backpack-laden hikers encounter a grouchy bear, they tend to act in predictable ways. One of those behaviors we can mention here is that humans drop their packs to gain those crucial few extra miles per hour. These bears discovered that those big sacks, which were left behind, carried good things to eat. Therefore, the grizzlies would just make a mean face, growl a few times, and take a menacing step toward a panicky hiker and Ba-Da-Bing! Free lunch. Good for the bears and bad for the hikers.

To break this cycle of experience, specialists played human voices over speakers (very loud speakers) to the bears, and then shot the approaching bears with the same rubber bullets that riot police use to break up rowdy, human protesters. They advised hikers not to drop their packs when charged by a grizzly bear—unless maybe that bear has a bruise from a rubber bullet!

We must strive to create an atmosphere where people will learn the right procedures, and keep the wrong ones in perspective. What if a firefighter breaks down a door without trying the knob? He has learned that he is powerful, that doors make cool sounds when hit with an axe, and that doors splinter when hit more than once. He believes he has accomplished his given task of gaining entry. True. True.

But is there an opportunity here for training and knowledge/ experience transfer? Definitely, *yes*. If an experienced firefighter had been with him, he could have stopped him and said, "Try the knob first." If the knob works, that information and example is branded on the new firefighter's mind and will allow him to use that information again at a later date. If the door is locked, the information is still branded on his mind but he also gets to destroy the door. *Value-Added Training!* Although this point of *try before you pry* has probably been emphasized in training, it does not seem to get transferred effectively until there is actual practice, application, and stress.

In a college psychology class, an instructor related a story about training and stress. Although an academic in the Navy, he was required to complete flight training. He was put through hours and hours of classroom training. Then the night before his flight, he was given an instruction book for the type of aircraft he would actually be flying. Being an academic, he had fairly advanced study skills and spent hours and hours that evening poring over the information, checklists, and sequences contained in the book.

The next morning he showed up at the airfield and was ushered to his aircraft by his trainer, a gruff, veteran aviator. After the pre-flight checks, he got into the cockpit and could not remember the procedure for starting the aircraft. Of course, the officer was not impressed, but transferred a precious gem for use in learning new material. He told the new pilot that if he wanted to learn something and make it stick, that he should review the material while bouncing a tennis ball.

What does tennis have to do with flying an aircraft? (It might depend on what kind of tennis player you are!) The simple stresses of concentrating on throwing and catching the ball makes the brain work harder to recognize, learn, and retain information. By doing this, information may take a little longer to plant in the recesses of your mind, but it will be there when you need it, too—especially under the stress of an emergency. So, although we transfer information every day in a non-stressful, classroom environment, our recall is not what we would hope when we arrive on the incident. By adding stress, even simple, safe stress, we can increase our absorption and recall of new material.

Using the stress of the job provides real life training, if we take the opportunity to use it. The point is that we pass up many opportunities to transfer valuable knowledge, experience, and techniques to our firefighters. On-scene training is not formalized or written. We do not stop the operation to bring everyone over to show them what an unlocked door knob looks like. But by supporting our people and supporting those little informal opportunities, we gain a lot.

The climate of *micro-training opportunities* is set prior to the incident, during the incident, and even after the incident. People who are proactive, motivated, and careful will take the opportunity to show and teach instead of standing back and shaking their heads (like those who are disillusioned do). Make sure that you set this climate in your briefings. Make sure people know that you expect this of them, you hope they do this, and you hope people being mentored will take the information in the manner it is intended, and not as a "Hey! Moron" example.

POSITIVE, PROACTIVE ATTITUDE

Just like the flu, a negative attitude has the ability to cripple your station's ability to do its job. Conversely, a positive attitude can be the energy source that propels you and your team to the top of its game. People with negative attitudes have no place on an emergency incident. An example of this occurred when an emergency medical technician (EMT) pulled up to a vehicle accident and took one look at the patient and said, "My God! That's the worst thing I've ever seen. I can't believe he's still alive!" What that patient was feeling then is what your firefighters will be feeling if you have a negative attitude. Your positive attitude, coupled with a proactive personality, will make your team successful and excited to be doing their job.

A positive attitude needs to be sincere. When you say how excited you are to be here, you cannot sound like comedian Steven Wright with his monotone, quiet voice, and asymptomatic expressions. Words do not matter as much as body language and actions. Conversely, it does not really energize the team when you are pathologically positive. If, at every scene, you walk around smiling and bouncy saying, "It's a good day to die," I bet people will not want to stand next to you.

Be positive about who you are, where you are going, and the job you are doing and people will generally join in—even if they do not always exhibit the same signs. People also like to follow proactive leaders who are grounded in reality at least a little and people who take personal initiative to make things better for themselves, their team, and the situation. People do not respect leaders who sit back and wait for things to come or to happen. The American Spirit lives on in the fire service where the pioneering spirit is the one that people are drawn to follow. Exemplify a positive and proactive leader in a briefing and people will also be willing and drawn to do the same.

Do not misunderstand the positive attitude. We should not be sugarcoating a bad situation. We can be positive during difficult times and in difficult situations.

A great motivational speaker described a meeting he attended when he worked for a large corporation. They were announcing that in a few months about half of those present would not be around due to company layoffs. As most people looked around the room and began to become quite negative and depressed, a voice from the back of the room spoke up and asked, "When all those people are gone...Can I get a bigger office?"[3] Was the situation bad? *Yes.* Was he, too, in danger of losing his job? *Yes.* Did he allow the situation to control his attitude? *No!*

The same applies for the emergency incident. Do not let a bad situation affect how you and your team view yourselves and the importance of what you are doing. This is important to maintain during the briefing, especially when things are going poorly.

Firefighters are usually most at risk for emotional problems when they feel helpless and are not doing something. Maintain a positive outlook on the situation and what the firefighters can be doing and you will, as they say in the West, head it off at the pass.

On many a wildland fire, I have been assigned to sit and do nothing. It almost seems like you do two things when fighting large wildland fires—wait for the fire and chase the fire. The crews, although bored stiff, do not allow the lack of work to create a negative attitude. They use the time for exercise, games, talking, etc.

It is very common to see a group of wildland firefighters standing in a circle playing *hacky-sack*. Of course, it has a firefighter spin on it because when you drop the hacky-sack or touch it with your hands, someone gets to give you a *hacky-bite*, which is what you get when they throw it at you from point blank range. Nevertheless, this is a method of creating and maintaining a positive attitude and climate under a very stressful waiting time for firefighters.

Never! Never, employ *busy-work* tactics in an attempt to, "Keep the crew busy." Using the *dig-a-hole-and-fill-it-up* mentality of past (and sometimes present), leaders can destroy morale, a team, and an individual. Firefighters' work should be meaningful and purposeful. Washing clean trucks and re-mopping floors that were just done last shift and haven't a speck of dirt on them will turn a good team into backstabbers and will destroy their trust and respect in their leader.

Tearing your team apart for no good reason, or assignment, is the mark of poor leadership. Leaders must choose the work for their team carefully. There are plenty of opportunities for team bonding and work if that is the goal of a leader. The death of a leader involves rigidly following a checklist or daily duties. Trucks that have not moved in weeks probably do not need to be washed daily.

In snowy or rainy weather, apparatus bay floors are doomed for disaster when the truck pulls in from the last call and daily washing commences. I have known crews who have spent most of their shift washing apparatus and bay floors after each call. This is an exercise in futility and frustration that centers on leadership with no common sense or flexibility to match the daily checklist and standards to the reality of the situation. Lazy leadership should be held accountable by the team and by the leadership of the department.

ACCOUNTABILITY

For every operation that goes well and every operation that goes poorly, there is always somebody who is accountable. CRM and team building appear on the surface to be contrary to the idea that people should be accountable for their actions/inactions. It just does not seem to fit that you could address substandard behaviors or skills and build your team at the same time. Well, that is exactly what you *can* and *must* do. We will discuss the actual debriefings or critiques in a later chapter and discuss how CRM thrives on this open, honest, and accountable atmosphere. For this section, suffice it to say that by remaining accountable for your and your team's actions and decisions, you can help to build a stronger and better team, and improve your safety and effectiveness over time. In the briefings, make it known who is accountable for what.

THE RISK OF OPERATIONAL ACTIVITY

For firefighters to be truly safe, to be injury- and death-free, we must never leave the comfort of our day room. Oh, I forgot. What about the risk of heart disease? Scratch that. We will only allow healthy firefighters in the day room; and, since we will not be doing any responding, we will not have to train as much (*read: at all*). Therefore, we will only hire firefighters between the ages of 22 and 28 and then kick them out and replace them with a new batch when they get too old. Then, *then*, we will truly be injury- and fatality-free. Think of it. No fingers pinched on extrication scenes. No smoke inhalation on fires. No contagious disease exposures. No training injuries. No on-scene deaths. Wow! What a wonderful, risk-free system.

Fortunately, that will never ever be the case. Even the safest buildings in the world need firefighters. They need someone to check the alarm system, preplan the building, inspect it for fire hazards, overhaul the small fire that was stopped by sprinklers, clear the smoke, and stop the flow of water. People will always need firefighters to take them out of wrecked cars, down the steps when they are too sick to walk by themselves, and to remove their children from folded hide-a-beds. Firefighters will never be out of a job. Our function, then, is not to dream up ways to put ourselves out of business, but to make our business safer for us and to provide a better service to our public.

We need to view our safety in a totally new light (maybe not so totally new for some out there). Safety is not a *part* of an operation, it is the operation. We do not do things *safely* we *safely* do things. We do not plan to go interior on a building and then say "Be careful in there. I don't know what that roof is doing." We should be looking at the roof before planning the interior operation. That is the difference between making safety *part* of the operation and making safety the operation. Too often, firefighters are caught up in the adrenaline of the scene and their *can-do* attitudes (or sometimes, *must-do* attitudes) and perform unsafe operations under the pretext of being on-scene, and therefore they must do something.

We must look at our operations, with regard to firefighter safety, by reviewing three points. The diagram in Figure 3–2 illustrates these points. The first and most important point is that we are always dealing with the unknown. We can never know all the hazards that are waiting to injure or kill our firefighters. With every operational action comes the risk of being injured by something unknown. "You don't know what you don't know." We can make efforts to reduce the weight of the unknown by implementing stronger building codes, inspection programs, and preplanning duties. Increased research into fire and its effects and increased training can also reduce this unknown factor. Although it is tempting not to address this unknown factor, we must do everything we can to reduce its size, discuss its presence, and plan for its surprises.

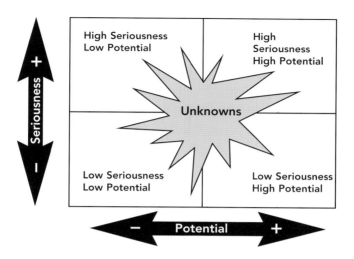

Fig. 3–2 Risk Analysis

The second point to consider is the likelihood that a problem will arise, or the potential that a problem will arise. Is the potential high or low or somewhere in-between? By allowing yourself to identify what hazards and problems might occur, you also look at the potential of these occurring. If you are fighting a wildland fire, there is a potential that a strong wind switch will bring the fire back on your position. At the initial stages of planning, the potential for this is unknown. But after gathering information on the weather patterns, weather forecasts, etc., you move the potential of a wind switch from unknown, to low or high depending on what you find.

We have addressed both items in one motion. We have identified that we do not know what the winds will do today. That is an unknown that we will keep in our minds when planning our operation and tactics. But we continued to gather more information and thereby moved the winds from unknown to low potential. Although we have moved the potential to low, we have not disregarded it completely; we still maintain our safety zones and escape routes.

The third point is the seriousness of the hazard or problem. On an emergency scene, there are literally thousands of things that can go wrong. Thankfully, most are very minor and are easily side-stepped without injury or major effect on the operation. When a firefighter looks at the seriousness of a hazard, he can put things in better perspective. If our hoseline bursts during an interior attack, it is very serious and plans must be made to reduce the seriousness of this occurring—usually a backup line is added. But if a hoseline bursts during mop-up/overhaul, it is not a big deal—the seriousness level is low.

Remember that this mission analysis and planning tool is directed toward safety and not operational effectiveness. If the hose bursts on interior attack, we look at how serious that would be for the safety of the firefighters and not that we couldn't put out the fire as quickly. Now let us look at an example of how to apply these items to standard operations.

Structural firefighting example

You are the first due engine company at a single-family dwelling. On arrival, at 14:35 hours, you see heavy smoke issuing from the rear of the single-story structure. Your second engine company is eight minutes out. You have four people on your engine and decide to mount an aggressive interior primary search. Due to the size of the fire, you opt to take a charged line in with you. With what you know now, plot some of the hazards/problems on the diagram. Were your actions appropriate for your mission analysis? (See Fig. 3–3.)

Wildland firefighting example

You and your engine are dispatched to a wildland fire in your area. It is August 12, 11:00 hours. The fire is approximately one-fourth of an acre and burning in a Fuel Type 7, light grass and brush understory with open timber overstory. It is burning near the top of an east-west ridge on an east aspect. The weather forecast calls for 89 degrees Fahrenheit, winds from the east initially, and then switching to the south with an increase of 5 MPH. You and your crew decide to hike to the lower edge of the fire and begin handline construction. With what you know now, plot some of the hazards/problems on the diagram. Did your mission analysis agree with your actions? (See Fig. 3–4.)

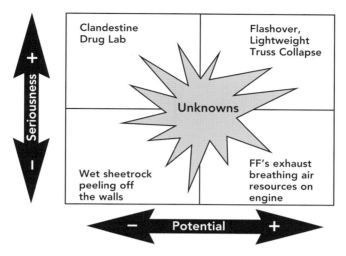

Fig. 3–3 Structural Firefighting Risk Analysis Example

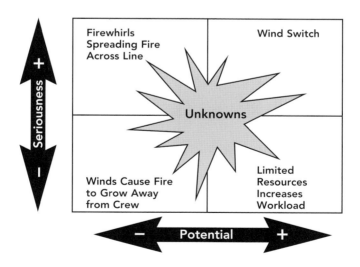

Fig. 3–4 Wildland Firefighting Risk Analysis Example

RISK-VERSUS-GAIN ANALYSIS

The most important thing to remember is that we should not trade lives for lives. For too many firefighters, there is an equation that identifies how many lives must be at risk for them to put themselves in a situation, which almost certainly means injury and possible death.

**My Life and Health are worth
more than Two Civilians
but
My Life and Health are worth
less than Three Civilians**

While we can agree that the unknowns we deal with are a factor in the number of firefighter injuries and deaths, we must also agree that firefighter risk analysis and risk acceptance are also direct and contributing factors.

The risk-versus-gain analysis is a very personal one. Looking at the same situation, most people will have differing opinions of what is the value we could gain by performing an operation and what risks we assume at the same time. There are many reasons for this difference in perceptions of the same problem. Experience, training, knowledge, trust, etc., are all parts of the analysis. How comfortable we are with our job performance, our protective equipment, past incidents, and the trust in our team members and leaders all have a major function in determining how we view the risk versus gain of an incident and operation. We might call this risk acceptance.

RISK ACCEPTANCE

The level of risk we will accept is a very personal process. It has to do with how all those things we mentioned before relate specifically with your job. But it also includes non-work-related information.

Cultural biases, gender, education, life experience, mental attitude, personality, and many other items figure into our thoughts of what level of risk we could or would accept. Most firefighters are generally not averse to accepting higher than normal levels of risk. The problem is not that we accept high levels of risk in our jobs, but that we are trained, equipped, and supported to operate at these unusual levels of risk. The problem is that we tend to exaggerate that expected level of high-risk acceptance. We tend to accept that high level plus a lot more—much of it unnecessarily.

"...if we wish to make an attempt at reducing [accidents/injuries]...[an] attempt should be aimed at reducing the level of risk accepted...."[4] Therefore, as firefighters and leaders, we must help firefighters adjust their levels of risk acceptance to a level that is both effective and safe.

For example, a fire department was dispatched to a multi-family, multi-story apartment building with smoke showing. Just prior to arrival, a firefighter on scene communicated that he had pulled a teenage girl out of the structure—she was badly burned. Upon arrival of the first unit, a neighbor reported that there were two young children inside—the teenager was babysitting them. Heavy smoke was showing from the apartment.

Due to the setback from the street and the second-story location of the fire, a hoseline needed to be built. A decision was made to conduct a dry search as quickly as possible—hopefully before flashover occurred. Two firefighters rushed up the stairs and entered the apartment. The smoke level was dropping fast, about two feet from the floor. Thermal conditions were very hot. The fire area was an open living room between the front door and the hallway to the bedrooms. A quick search of the kitchen and living room area was mostly visual due to the heat encountered.

As the firefighters were progressing down the hallway, a flashover occurred in the living room. A couch, two chairs, and the carpet ignited. Instantly, heat levels increased intolerably. Carpet in the hallway was melting onto the firefighters' gloves and knees; helmet visors drooped, and bunker gear was scorched. Just then the firefighters received a communication that the children were safe outside. The firefighters quickly retreated and attacked the fire with the hoseline that had just been stretched to the door. All in all, the result was one serious burn injury to the teenage babysitter, some ruined firefighting equipment, and lots of new knowledge and experience. To apply the risk versus gain analysis to this incident is easy.

Upon arrival and application of aggressive, dry, primary search:

Possible Gains

- Two small children found

- Fire extension to other apartments and areas slowed

Possible Risks

- Flashover conditions

- No immediate water stream protection was available

- Firefighter injury or death could occur

After new information was received and crews moved to attack the fire:

Possible Gains

- Stopped fire extension

- Changed strategy and tactics increased firefighter safety

Possible Risks

- Presence of high thermal levels

- Bad smoke conditions

As you can see, the firefighters in this real-life example initially accepted a huge risk for rather large gains. But as new information was introduced, not one person had a problem with the application of a new strategy for fighting the fire because of the high levels of risk and the low levels of gain.

The lessons learned here are that we need to accept the level of *risk* that

- We can accept without injury

- We can correlate with the expected gains

- We can support

The lessons learned here in regards to the level of *gains* are

- We should not trade lives for gains—we are not the military

- We must continually adjust our gain expectations to the reality of the situation

Many times, our risk-versus-gain analysis takes on a silver-lining view that looks only at the good things and not the bad. We look at the outcomes we would expect if we had perfect conditions, the perfect fire behavior and location, the perfect partner, no problems, etc. What often happens on emergency incidents is just the opposite.

In response to a grass fire call, the first officer on-scene identified the need for additional resources. The grass fire had spread to a few structures, many vehicles, trash piles, lumber stockpiles, etc. When I arrived on-scene of what I called the, "Y'all-Come-Fire," the officer presented us with a menu of fires to choose from—this reminded me of a drive-up menu at a local fast food joint. Our crew was assigned to a couple of car fires. Three of us had stretched a $1\frac{1}{2}$" line to a vehicle that was burning. We doused the engine compartment briefly, then opened the passenger door to perform a primary search, and extinguish the fire extension in the passenger compartment. Just as the door was opened, we heard (almost felt) a loud *pop!* The tire on our side had blown.

I suddenly had trouble holding the hoseline. I shut down the nozzle and looked back to find that the other two firefighters were *gone!* Both firefighters were rookies and had concluded that the tire blowing had spelled their doom and they had decided to die on the run. This example illustrates that the risk acceptance that I had and the level that they had were not the same for various reasons. They did not have experience on many vehicle fires and therefore did not know what a tire blowing sounded like. I knew and accepted that small risk. They did not. Your risk analysis must be tied to the risk analysis and acceptance levels of your team members; otherwise, you could be in for big trouble. By the way, we picked up the two rookies halfway back to town—still running!

In our risk-versus-gain analysis of our incidents and operations, remember

1. Everyone's perceptions of the risks and gains are personal; and, therefore, unpredictable, hard to explain, and unknown to others.

2. Some risks are too great no matter the expected gain.

3. Continually adjust the risk-versus-gain analysis to match the situation.

4. There is a fine line between too much risk and not enough risk. Always err on the side of less risk if you are unsure of yourself, your partner, or your team.

THE CHANGING OPERATIONAL SETTING

We all know that even the best plan does not usually last very long on a dynamic emergency scene. Good teams always have a plan, a backup plan, and are working on the backup-backup plan.

Maintaining an appropriate balance between planned and flexible response seems to be a key requirement for military effectiveness:

creativity and flexibility, within a cohesive framework of planning and preparation. Immersive mission planning and rehearsal, and assertive mission automation, should be undertaken with caution. In particular, care is needed in making assumptions concerning the predictability of real-world events, and to assist in real-time mission re-planning....In an uncertain world, all plans are inherently fallible.[5]

So, in our briefings of what we think is happening, what the hazards are (seriousness/potential/unknowns), and what our plan is, we need to maintain our vigilance for the changing situation and the necessity for a changing plan. A good briefing is spelled out in this reference:

A superior crew mission briefing, in both content and context terms, was demonstrated by the most effective crews. They included a broad range of planning factors within their pre-mission briefings, and they developed high quality...plans. Ineffective crews discounted the importance of mission planning, and performed their pre-mission briefings even before an execution plan had been adequately prepared.

The most effective crews developed aggressive execution plans, with a high level of consideration given to development of mission options and divert plans. Additionally, the effective crews showed an overt awareness of time as a finite resource, and they were often observed monitoring and questioning the time management status throughout the planning, preparation, and execution phases.[6]

The bottom line is that crews perform better when given more information. This information is usually contained in the IC's head, but not usually communicated to the people on the ground. This gap leads to delays in communicating plan changes, delivering the appropriate information to the IC, and hazard communication. By showing everyone the playbook, the IC points out the current plan, the hazard expectations, what communications are important from the field, and what to expect in a backup plan. Highly effective crews get a good briefing that allows them to deliver the pertinent information. This information then allows the IC to monitor the progress of the plan and operation, and to adjust "on-the-fly," rather than stopping, requesting information, re-adjusting, and re-communicating.

Give a good briefing; address the risk-versus-gain analysis; remain flexible to real-world situations; and your crew will be safe and effective.

REFERENCES

1. Helmreich, R.L., H.C.Foushee, R. Benseon, and R. Russini, "Cockpit Management Attitudes: Exploring the attitude-performance linkage." *Aviation, Space and Environmental Medicine*, 57, (1986): 1198–2000.

2. Adapted from: *United States Coast Guard. Effective Mission* Analysis. *http://www.dirauxwest.org/effective_mission_analysis3.htm*

3. Keith Harrell, Motivational Speaker. Fire West Conference Notes.

4. Gerald J. S. Wilde, Ph.D., *Target Risk 2: A new psychology of safety and health. What works? What doesn't? And why....* PDE Publications, 2001. 310-5334 Yonge Street. Toronto, Ontario, Canada M2n 6M2. Author Address: Professor Emeritus of Psychology. Queen's University. Kingston, Ontario, Canada K7L 3N6. *wildeg@psyc.queensu.ca* :6.

5. Taylor, R.M., S. Finnie, and C. Hoy, "Mission Planning: Cognitive Rigidity: The Effects of mission planning and automation on cognitive control in dynamic situations." In Jensen, R. *Ninth International Symposium on Aviation Psychology*, Columbus: Ohio State University, 1997.

6. Tourville, S., "Identification of key behaviors for effective team coordination in special operations forces combat mission training." In Jensen, R. *Ninth International Symposium on Aviation Psychology*, Columbus: Ohio State University, 1997.

SITUATIONAL AWARENESS

4

The Killer Equation—Reality times Perception still equals Reality.

WHAT IS SITUATIONAL AWARENESS?

SITUATIONAL AWARENESS is the skill of becoming aware of the situation as it actually exists. Usually there is a huge difference between how someone perceives the situation and how it actually exists. There is also a difference between how a person analyzes a situation and how the actual situation exists. We all know the story about the blind men and the elephant. One blind man felt the leg, and thought the elephant was a tree. One felt the tail, and thought the elephant was like a bush. One felt the ear, and thought the elephant was like a tent. One felt the tusk, and though the elephant was like a tree limb. All were correct. All were wrong. Situational awareness training teaches the skills necessary to use resources to determine how the situation actually exists so we see the whole elephant. More importantly, situational awareness training teaches the signs and symptoms of the loss of situational awareness.

How Do We Train for Situational Awareness?

The firefighting community does a good job of teaching its firefighters which dangerous situations to look for on the fireground. Firefighters are given extensive training on the nature of fire, building construction, and the risks on the fireground. Before they are allowed to pick up their first Pulaski, a specialized wildland firefighting tool which is half ax and half hoe, wildland firefighters are trained on fuel types, weather, and fire behavior.

Before we ever allow firefighters on the line, we give them extensive training on what dangerous conditions exist. How often, though, do we train them on when to look for those situations? How often do we place firefighters in realistic stressful situations, and teach them how to maintain their situational awareness?

Good situational awareness is the next best thing to a crystal ball. To fight a fire, you need to be aware of three things: the *fire*, the *plan*, and the *people*. We need to be aware of what is happening right now and what will happen in the future. If we can maintain good awareness of the present situation and we have good knowledge of the anticipated behaviors, we can accurately project the situation into the future.

The skills necessary to maintain situational awareness include monitoring, evaluating, anticipating, and considering.[1] The first skill is monitoring. Put very simply, monitoring is being aware of everything that you need to think about and ignoring everything else. Monitoring means fully using all of your senses as well as the senses of every other member of your crew to gather all of the relevant data for your situation. Selection of relevant information is, of course, a product of experience and training and will be discussed more fully in the chapters about decision-making.

The next step is evaluation of the information. It is not only important to know what you are seeing but to know why what you are seeing is important. For example, if you are a wildland firefighter, and you see dark cumulus clouds building in the west early in the afternoon, it does not mean that you will be lying on your back looking at fuzzy bunnies in the sky. What it means, to the trained eye, is that your crew should anticipate a storm front passage and wind shift, both in direction and intensity, later in the day. The evaluation of relevant information is also a product of experience and training.

The next step in maintaining situational awareness is anticipating developments. If you are acting on the information you have now, your decisions are obsolete when implemented. Time passes between the time the decision is made and the time the actions are implemented. By the time the actions are implemented, the situation has changed and the action may not be appropriate for the situation as it exists when implemented. Consequently, thinking ahead of the fire is the key to maintaining present situational awareness.

The final step in maintenance of situational awareness is to consider contingencies. If situation X happens, consider evolution Y; this is the equation for success. We constantly train in automobile extrications for command to be one or two evolutions ahead of the people working the extrication. If plan A does not work, switch to plan B.

By thinking ahead of the situation and considering options before they occur, we avoid the time stress of decision-making. When the contingency happens, the option is already developed. Better decisions can be made more quickly because we are ahead of the situation and not trying to catch up with the rapidly unfolding events.

WHEN DO WE LOSE
SITUATIONAL AWARENESS?

Firefighter CRM training should include not only the dangerous situations firefighters should avoid but also the clues that point to the loss of situational awareness. Factors like complacency, high stress level, ambiguous instructions, unresolved discrepancies, lack of experience, lack of communication or coordination, fatigue, lack of adequate weather information, emotional pressure, fixation, and just a bad gut feeling are clues that situational awareness is being lost.

The firefighter should be taught that when these elements start to arise, it is time to take a step back and evaluate what the situation is and what the plan will be if things start to go wrong. Periodically throughout any operation, the firefighter should ask the following:

- Am I aware of what is going on around me?

- Are things happening as they are supposed to be happening? If not, why not?

- If things should go wrong, what is the plan?

- Does the leader know the answers to all of these questions?

If the answers to the previous questions are unsatisfactory, all firefighters should be given the authority to completely stop any operation in which they are participating until satisfactory answers to the questions can be given. To maintain situational awareness, the most important thing upon which to focus is, *what* is right, not *who* is right. Crews need to strip themselves of rank, bias, prejudice, cultural, and communication issues, and focus on the situation at hand.

CLUES TO LOSS OF SITUATIONAL AWARENESS[2]

Clues exist, which lead one to be suspicious of a loss of situational awareness. Use of these clues can give the firefighter a series of warning signs to reevaluate the situation and make sure the *theory of the situation* matches the actual situation. Those clues for loss of situational awareness include the following.

Fixation

Fixation or tunnel vision is one of the most common incidents on a fire scene. The emergency responder is faced with a dire situation, and the focus naturally narrows down to the most serious situation facing the emergency response team. It is also one of those situations that kills firefighters. How often has a ladder contacted a power line, killing the firefighter holding the ladder? How often have firefighters been injured in a structural collapse because they were fighting the fire too long, and the building just gave way?

Basic firefighter training always tells us to look up before putting a ladder up. Any firefighter who has had basic building construction training knows the building is going to fall down, if it has been on fire for a certain period of time.

One of the very first incidents that led to the development of CRM, was the crash of an Eastern Airlines jet into the Florida Everglades. All of the people in the cockpit were focused on a burnt out light bulb, and no one was flying the plane. Basic pilot training teaches all pilots to fly the plane. The highly experienced pilots of that airplane forgot.

The fire service has developed several strategies for avoiding fixation. It is when we vary from these strategies that we develop tunnel vision and we get hurt. The first step we take to avoid tunnel vision is to make sure that the command officer remains at the command unit. The command officer is a highly experienced firefighter whose sole job on the fire scene is to maintain an overview of the fire. When the command officer is at the end of a hose line, there is no way to maintain an overall view of the fire. If the command officer is pumping an engine, the command officer is not watching fire behavior. When the command officer is no longer at the command unit, that fact is a clue to all firefighters on scene that overall situational awareness is being lost, and every firefighter should take extra precautions to understand the situation.

The second thing the fire service does to avoid fixation is to appoint a safety officer, whose sole purpose on the scene is to ensure all operations are conducted safely. The safety officer is not responsible for throwing ladders. The safety officer is responsible for looking for power lines that can be contacted by ladders; the safety officer is not responsible for interior attack; the safety officer is responsible for observing signs of building collapse. When the safety officer is on a hose line, throwing a ladder, or rendering first aid, all firefighters on the scene have a clue that overall situational awareness is being lost.

The third thing the fire service does to avoid fixation is to provide time prompts from dispatch or an on-scene timer. Those prompts serve as a reminder for the command officer, the safety officer, the division officers, and the firefighters to take a moment and observe the situation around them. When the reminder tone goes off, each firefighter should think, "Hey, what's really going on here? What is the fire doing? Am I obtaining good results, or do I need to readjust my strategy?" When the on-scene prompts are not being used, it is another sign that overall situational awareness is being lost.

It is a natural tendency in a stressful situation for the human animal to develop tunnel vision. We, as firefighters, need to counteract this natural tendency with artificial devices. To ask a human being not to develop tunnel vision is like asking a human being not to breathe. We have evolved those artificial devices by stationing the command officer in the command unit, by creating a safety officer position whose job it is to augment and counterbalance the command officer, and by using artificial timing devices that prompt us to remember where we are, and what we are really doing. When we utilize the tools we have developed for ourselves, we can avoid tunnel vision.

How often have firefighters been hurt because the firefighters were concentrating on interior suppression while no one was timing how long the building was on fire, resulting in a building collapse on top of firefighters after the predetermined interior time of an uncontrolled structure fire had been exceeded? We must remember three things:

1. Fire—What is the fire doing?

2. Plan—How is the plan working?

3. People—How are the people doing?

Situational awareness is the correct perception of the situation as it actually exists. The "theory of the situation" is what a person assumes to be true for a specific period of time. If what a person assumes to be true is different than what is actually true, then a loss of situational awareness occurs, and the dominos of a disaster can begin to fall.

Overconfidence

A factor closely related to fixation that causes a loss of situational awareness is overconfidence. The belief that one knows exactly what is happening and is in absolute control of the situation is a type of fixation that lulls the firefighter into a sense of false security. In addition to the tools available to combat fixation, a firefighter can play the "what if" game.

The aviation industry constantly trains its pilots to anticipate problems and to address them in a systematic way. They are trained in a simulator on the proper procedure if the #3 engine catches on fire. Then they are asked while in flight to rehearse in their minds what would happen if the #3 engine catches on fire.

Similarly, firefighters train for unanticipated consequences. For example, they are taught the hazards of car fires. They are told that overheated air-charged bumpers could become projectiles. They are taught that car fire smoke is some of the most toxic smoke a firefighter faces. After being taught these hazards, they are taught to think of steps that can be taken to avoid the hazards.

That training can be extended to the emergency scene, to help maintain situational awareness. While on scene, a firefighter can pose hypothetical "what-if" situations, and then solve the problem. For example, a firefighter might ask at a routine car fire, "What if the trunk is full of explosive chemicals? What would I do?"

By posing on-scene what-ifs, the firefighter's mind is disciplined to look for the unexpected. As discussed later in the decision making chapter, the posing and solving of what-ifs creates patterns that will aid decision making at future incidents.

Distraction

The next factor that is an indicator of loss of situational awareness is that of distraction. Information overload was recognized as one of the problems facing firefighters by the Tri-Data study.[3] At an emergency scene, many things are happening all at once. The demands on the firefighter and command officer are to assess the scene, determine any dangers, develop a plan for mitigation of those dangers, and then conduct the planned mitigation.

However, on the emergency scene, many things pull the fire officer in different directions. Victims are screaming in pain, units are looking for assignments, the building needs a size up, the news media wants a comment, and other agencies (such as law enforcement) are looking for coordination. All of these factors work to shift focus from the task at hand.

In addition to the on-scene distractions, our minds contain a series of prejudices, emotions, filters, and biases that we bring to the fire scene. A firefighter brings to the scene a surprising array of preconceived distractions. The firefighter may hate one of the members on his crew for a bad financial transaction. The firefighter may have cultural difficulties with other people on the crew or at the emergency scene. The firefighter may come to the scene upset because of a fight with a spouse over which daycare to enroll a child in. One of the victims may remind a firefighter of a family member—particularly a child victim. All of these distractions can serve to derail situational awareness.

The first step in avoiding a loss of situational awareness caused by distraction is to minimize the level of distractions. The aviation industry has developed the "sterile cockpit" concept. In the sterile cockpit, communications are limited to the task at hand during certain phases of a flight. Conversations about what happened last night or what the crew had for dinner are specifically prohibited during critical phases of flight, such as takeoff or landing. The fire service can apply this concept. Any time a piece of apparatus is running code, or while rescue or initial suppression operations are ongoing, conversations should be limited to the task at hand. In taking this simple step, critical distractions can be avoided.

The second step in preventing a loss of situational awareness caused by distraction is to train to recognize the distractions. Simply placing a video camera in the command unit, focused on the command officer will give individual departments an amazing insight into the workload placed on the command officer. The command officer in many departments is obligated to drive the command unit, talk on the radio, answer the cell phone, write down critical information, formulate a plan, and read a map to the scene all at once. Spreading attention amongst so many critical tasks is a recipe for disaster, but is more often than not a common practice. Reviewing the distractions each individual department places on its firefighters and officers provides the first step in limiting those distractions.

Finally, the last step a department should take in avoiding a loss of situational awareness as a result of distraction is to develop a series of standardized protocols and checklists. If communications are standardized, less time is required to convey more information. If the department approaches similar events with the same protocol every time, there are fewer chances for deviations.

The aviation industry requires pilots to follow checklists for every operation they perform. Pilots are prohibited from memorizing the checklists, and are required to read them every time. The copilot acts as a check to insure every item on the list is completed. While the nature of the fire service does not lend itself to a complete set of lists, a series of checklists followed by firefighters and command officers alike would standardize the operation. Doing the same thing the same way every time results in matched expectations and fewer distractions.

Information overload

A factor related to distraction is information overload. Information overload was recognized as a problem for wildland firefighters by the Tri-Data study.[4] The skills of prioritizing and filtering information must be developed among firefighters. In addition to prioritizing and filtering information, firefighters need to be trained in the dangers of flooding people with too much information. Timely information is needed to keep current on the situation, but information overload can be as devastating as not having enough information. The more experienced a person is, the better their ability to assess the situation. Consequently, less raw information is needed to make command decisions. What a relatively inexperienced person might see as 10 pieces of information, the experienced person might see as one or two.

The human animal reacts to stress in a predictable fashion. As information overload increases, short-term memory decreases. The human animal begins to rely on those memories stored in long-term memory. For the experienced firefighter, those memories include memories of reactions in similar situations that worked. Behaviors that saved a firefighter in a dangerous situation before are resurrected and used in the new dangerous situation. As long as the prior learned skills are fundamentally sound, the firefighter calls upon those previous memories and uses them to survive the situation at hand.

For the inexperienced firefighter, those memories do not exist. All the firefighter has to rely upon is what occurred in training. There are no long-term memories upon which to rely, and the systematic thought process is frozen. The firefighter freezes and becomes the proverbial "deer in the headlights." As stress increases, we cannot expect our firefighters to think their way out of a stressful situation (although they often do). We need to realize that we are dealing with a human animal, and there are times when, through no weakness other than just being human, they become overloaded and freeze. We need to train our firefighters to react their way out of stressful and dangerous situations. We cannot overload a pump without it cavitating, and similarly we cannot overload a firefighter's mental process without them freezing up.

As experience levels increase, the ability to process relevant information also increases. In order to reduce information overload, the fire service must work to increase the experience level of firefighters, reduce the number of orders and rules, and train the people to use the radio efficiently. Decision making under stress is critical to success on the fire ground, and is further developed in another unit of the CRM program.

One strategy for training to avoid information overload is to "over-learn" a particular skill. Over-learning is defined as the deliberate process of over-training a task beyond the level of initial proficiency. One study found that a group that over-learned a particular task made 65% fewer errors than a group that did not over-train on the same task, when tested later.[5]

A strategy for decreasing stress in the less experienced firefighter is to pre-brief the situation. Giving an individual as much knowledge and understanding as possible regarding future events reduces stress.[6] If we can give the firefighter a look at what to expect and why, we can reduce the stress levels and increase the levels of situational awareness.

Communication

Communication is such an important topic that we have devoted an entire chapter to the subject. However, we would be remiss in not discussing communication in a situational awareness setting.

It is nearly axiomatic that poor communication skills lead to a loss of situational awareness. If two parties cannot communicate critical information to each other, by definition there is a loss of situational awareness.

On the other hand, if communication is done effectively, team performance is enhanced. The object of communication skills is to develop a shared mental model of the situation as it actually exists. A shared mental model is simply for all members of the crew to have the same mental picture of what is happening.

For example, when a command officer at an automobile accident tells a rescue crew that he would like a roof removal, each member of that crew has a picture of exactly what the command officer wants. That shared mental model allows the crew to perform in harmony. Now, if that same officer asked his crew to perform the same tasks by giving the following set of commands, the picture would not be as clear: "First remove the windshield, then cut the A post, B post and C post on each side of the car. Then, remove the roof and discard it."

Those crews that have developed the shared mental model consistently outperform those crews that do not develop such a model.[7] The reason the shared mental model allows better performance is that each member of the crew can anticipate what other members of the crew are doing, anticipate the workload of other crew members and work to spread the load more evenly, and share the burden for looking for potential hazards in the operation. A shared mental model can be developed through basic training in team skills and then practicing those skills in a stressful environment.[8]

The purpose of effective communication in a situational awareness context is to utilize all of the combined senses and experiences of all of the people at the incident to gain and share an accurate picture of what is really happening on scene. Through effective utilization of all on-scene resources, awareness of the situation can be maintained.

Low stress level

When we are not on our guard, we are not on our guard. In other words, if we are not looking to prevent bad things from happening, bad things happen. Our history as a fire service has shown that we most often get hurt in the least dangerous parts of our operations.

- Most incidents happen on the smaller fires or isolated portions of larger fires.

- Most fires are innocent in appearance before unexpected shifts in wind direction and/or speed results in "flare ups" or "extreme fire behavior." In some cases, tragedies occur in the mop-up stage.

- Flare-ups generally occur in deceptively light fuels such as grass and light brush.[9]

Complacency is one of the major keys to loss of situational awareness on the fireline. A fireground injury or death is just like a trip into a horror movie, and just like any self-respecting horror movie, when you don't expect it—*expect it*.

High stress level

When your heart is pounding so hard you can hear it in your ears, you really don't know what is happening around you. Your fight or flight instinct is beginning to kick in, and you are looking for a major threat. You are not taking the time to figure out what is happening around you, and as a result, you are not situationally aware.

Chief Gary Scott of the Campbell County Fire Department in Gillette, Wyoming, teaches that firefighters should be aware of their heart rate index. With no exertion, if the heart is beating at 60 to 70 beats per minute, the firefighter is not paying attention to what is going on. If the firefighter's heart is beating 70 to 90 beats per minute, the firefighter has the appropriate level of stress, and is concentrating on what is happening. If the firefighter's heart is beating between 90 and 110 beats per minute, the firefighter needs to take a few deep breaths and think about what is going on. If, with no exertion, the firefighter's heart is beating over 110 beats per minute, it is time to take a few steps back, calm down, and get back into the game. An overstressed firefighter is a firefighter who is not aware of the situation around him.

One of the strategies for effective performance in high stress situations is to train in situations similar to those likely to be encountered in real world situations.[10] In other words, the fire service needs to practice like it plays. The aviation industry has spent millions of dollars in development of high fidelity simulators for pilots to practice for eventualities they may never face. The fire service needs to develop and implement high fidelity training devices that allow for high stress training in realistic, but safe, environments.

Lack of experience

Lack of experience comes in two varieties, personal and situational. A rookie firefighter lacks personal experience. The rookie has been to limited emergency scenes and consequently has a limited view on what to expect and anticipate. The inability to know what indicators are important and what indicators are not important contributes to a lack of overall awareness.

On the other hand, situational inexperience can affect everyone. Firefighters are called to address dynamic situations in unfamiliar territory. Being unfamiliar with local conditions, terrain, weather, building construction, hazardous materials, or local cultural customs can be a recipe for disaster. Firefighters need to familiarize themselves with all available information before, during, and after an incident to build on their knowledge and understanding.

Fatigue/illness

When we remove two spark plug wires from an eight-cylinder engine, we don't expect the engine to run right. When we shut down one engine on a three-engine airplane, we don't expect the airplane to fly at full efficiency. When we pump water through a one-inch line instead of a deuce and a half line, we don't expect to flow two hundred and fifty gallons of water a minute. Why then, as a fire service, do we often make the mistake of working our people long hours, or when they are not feeling well, and then expect them to perform perfectly?

A tragic example of the effects of fatigue on firefighters is the 2001 Thirty-Mile Fire at Winthrop Washington, in which four firefighters were killed in a burn over. For the first time in a major fire investigation, a human factor fatigue analysis was included. The analysis of the human factors of the event concluded there were clear indicators of low situational awareness, possibly exacerbated by sensory or perceptual factors.[11] The report goes on to say that the single overwhelming physiological factor that impacted upon this mishap was fatigue caused by sleep deprivation. The fatigue experienced by the firefighters, the report indicated, may help explain the "uncharacteristic lapses in judgment and the multiple violations of the 10 standard Fire Orders and the 18 situations that shout Watch Out."[12]

If a firefighter feels fatigue, or observes fatigue in any other members of the team, that fatigue is an indicator of a loss of situational awareness.

Reliance on machines

Just because a machine says it is so, it may not be so. As technology increases, and our reliance on machines grows, we have to understand that machines break. Reliance solely on mechanical indicators may lead to a loss of situational awareness.

For example, a HAZMAT team responded to a butane truck rollover. The butane tank was ruptured and ice was forming on the ground below the hole. The initial responders took a combustible gas indicator around the scene to determine the size of the combustible gas plume. The gas indicator did not indicate any explosive gasses. The initial responders could smell the mercaptan in the gas, but relied on the machine to the exclusion of their senses, thus placing themselves in an unnecessarily risky situation. When the HAZMAT team arrived on scene, they quoted the old adage, "If you can smell it, you are in it."

If the machines conflict with your senses, resolve the conflict. Don't discount your senses just because a machine says you are wrong.

Unresolved discrepancies

Relying on machines when our senses tell us something different is an example of an unresolved discrepancy. When two pieces of information do not match, something is happening that needs to be investigated.

Unfortunately, human beings tend to want to change the facts to meet the theory of the situation, then to change the theory of the situation to meet the facts.[13] When a crew wants to change observed facts to mold the perception of the situation, situational awareness is completely lost. When a firefighter observes facts being changed to meet theories, it is time to take a step back and resolve unresolved discrepancies.

Professional attitude

Professionalism is a matter of attitude. The attitude and culture that presently exists in the fire service is a can-do attitude. While that can-do attitude is important to accomplishing difficult tasks, the fire service must modify the attitude to a can-do-safely attitude. There is no way to "just say no" in the fire service that does not carry formal or informal sanctions. We are advocating a change in our culture. We suggest that a professional firefighter is one who operates in a culture of safety, where risks are assessed, weighed, and actions determined based on all the information. Our goal is that all firefighters go home to their families after the call is over.

Looking for ghosts

One of the most important indicators of situational awareness is to look for what is not there that should be there. If something is missing, a little situational awareness alarm bell should ring. In this time of terrorism, this indicator is particularly important.

A common example of the thing that is not there is the child abuse EMS call. The child shows profound injuries, but the mechanism of injury does not match the story being told by the parent. The missing appropriate mechanism of injury should cause the little bells to ring.

Gut feeling

If it doesn't feel right, it isn't right. "Our bodies are able to detect stimuli long before we have consciously put the big picture together."[14] Through experience firefighters must learn to recognize their own signs such as stomach butterflies, muscle tension, mood swings, or inability to get along. Trust your feelings; firefighters place their lives on gut feelings.

One of the most compelling examples of failure to rely on "gut feeling" is the Storm King Mountain Fire near Glenwood Springs, Colorado, in 1994. Fourteen firefighters died on that mountain. However, discussions amongst the firefighters earlier that day revolved around that piece of ground not being worth dying for. The firefighters sensed something was wrong, but could not articulate the specifics.

If your body is telling you something is wrong, it is important to perk up and listen to what your body is trying to tell you. Your gut level perceptions do not suffer from mental filters, prejudices, or processing. They just are. And when your feelings tell you something, your life may depend upon you listening.

The "Ah pooh!" experience

The nature of firefighting is that firefighters are summoned to emergencies that range far beyond fires. Whenever the public is faced with a situation it has no experience handling, they call the fire department. Consequently, firefighters are summoned to every type of emergency situation that can be created by man. Unfortunately, man has demonstrated an extraordinary ability to create infinitely variable emergency situations. So, as adept and well trained as a fire department is, invariably it is called to an emergency scenario far beyond the realm of experience of any person on the department, or for that matter beyond the experience of anyone else in the world.

The "Ah pooh!" scene gets its name after the first two words frequently uttered (or exclaimed) by the first firefighter on scene. Well known examples of the "Ah pooh!" experience include the World Trade Center on September 11, 2001, the Oklahoma City Bombing (1995), the San Francisco Earthquake of 1906, and the massive burning of the wildlands of the United States in 1910. "Ah pooh!" incidents, however, are not always spectacular, headline-grabbing events. There are many daily "Ah pooh!" experiences that never make the press. All firefighters with any seasoning have an "Ah pooh!" story in their repertoire.

An example of an "Ah pooh!" event in the experience of the author is: A vehicle collision involving a tractor trailer carrying 7000 pounds of propane collided with a tractor trailer hauling 3000 gallons of gasoline. The gasoline trailer was pulling a pup with 1500 additional gallons of gasoline. The vehicles left the road and pushed through a fence, coming to rest next to a major power line crossing a pasture. The pasture was carpeted with tinder-dry 3-foot tall grass on a 105 °F day. Rounding out the "Ah pooh!" for the arriving firefighters was the herd of very ill-tempered bison that were quite disturbed at all the commotion interrupting their grazing.

Training for the "Ah pooh!" experience includes first, saying the two magic words that describe both the bodily function that the firefighter is experiencing and the emotions felt in observing the situation. After the two magic words are uttered, the firefighters are trained that the mere utterance of those words indicates a need for the extreme application of situational awareness skills. Next, the firefighters are trained that they did not create this situation, and the worst thing they can do is to charge in and make the situation worse.

The next step taught is to break the situation into manageable tasks rather than being overwhelmed by the scope of the entire task. Tasks are then prioritized under the mantra, "Address what is going to kill you first, and then move to the next threat." The firefighters are trained to seek out whatever help and expertise is available, but to move in with a clear assessment of the dangers and a clear plan for preserving life and savable property.

"Ah pooh!" experiences can be so overwhelming that even obvious steps that can minimize risks are not taken. If the firefighters are not trained to address the experience in a systematic way, the reaction can degenerate into a series of knee-jerk responses without assessment of the consequences. These knee-jerk reactions typically result in death and injury.

There is a need in the fire service for an assessment of firefighter actions and reactions to the risks that occur during the "Ah pooh!" experience. The assessment would give rise to developing a series of steps that could be taken to address those risks in a systematic way. Part of the assessment and analysis would include evaluating the thought process of firefighters who faced the "Ah pooh!" situation and handled it with positive outcomes. Right now the training, when it does occur, is based upon firefighter intuition and experience without any scientific proof to validate the training scenario.

DEVELOPING TOOLS TO MAINTAIN SITUATIONAL AWARENESS

Just like with any other task, having the correct tools, properly aligned, facilitates situational awareness. There are things the individual firefighter can do to prepare for maintaining situational awareness. Here are some suggestions to consider.

Experience

There is no substitute for experience. Experience creates a mental file. Firefighters draw upon their experience every call, and use it to assess conditions and make decisions. Under pressure, people tend to revert to previous patterns of behavior. An individual's experience file establishes how one will interpret and respond to a given set of conditions. Experience is not a destination. You must always be learning and watching. Firefighters who make decisions based solely on experience are edging toward oblivion.

Training

Training does far more than perfect skills. Training adds to a firefighter's experience by creating events that rarely happen in real life.[15] Training sessions can generate a lifetime of experience in a very short time. Few firefighters will ever see a blow-up resulting in fire shelter deployment or the onset of a flashover, yet through training, those experiences can become part of a firefighter's experience. So if and when they do occur, there will be something to draw upon.

Training must stress repetition of tactics to be used in emergency situations. We know that the human being resorts to long-term memory to guide actions in stressful situations. If we rely on people's ability to think in a stressful situation, we are training them to fail. For example, deployment of a fire shelter should be an automatic action. Repetitive training of shelter deployment should result in a reflex action, rather than a step-by-step action.

Firefighting skills

Firefighters still have an obligation to fight fire efficiently and safely. Training should stress basic firefighting tactics. Creation of good firefighting habits will pay off because in a stressful situation, firefighters will revert back to their training and experience. Firefighters must have a solid skill base before they can become effective team members.

Ability to process information

The ability to use information from sensory organ inputs, observation, and other sources to form an accurate picture of what is happening is a skill that takes practice to use under stress. It is the ability to integrate all of the elements that contribute to situational awareness.

Professional attitude

A professional attitude that stresses safety over accomplishment does not just happen. It is a conscious effort. To be safe, one must think safe. There is a difference between going out and doing the job and going out and doing the job right. The attitude we carry with us to the job determines the success or failure of the job. If we train professionally and act professionally at routine scenes, professionalism becomes automatic at times when we need it most. Simple actions should be taken seriously.

Emotional/physical condition

An individual's emotional and physical condition affects perception of the environment. Firefighters with poor emotional or physical conditions should consider sitting out a call until the condition can be remedied. Fifty percent of firefighter deaths are directly related to physical condition. Good physical and emotional health means firefighter safety.

A STRATEGY FOR MAINTAINING SITUATIONAL AWARENESS

Situational awareness is a process. It begins with maintaining control, assessing the problem in the time available, gathering information from all sources, and monitoring the results of any action. If the firefighter constantly follows this process, situational awareness can be maintained.

Maintain control

The first step in maintaining situational awareness is to maintain control. Even though you have a problem that needs to be solved, you must remember to keep a global view of the situation. On a recent car fire the pump failed to engage. The Power Take Off (PTO) valve was stuck and the pump was not going to pump water. The entire engine crew became focused on repairing the pump instead of fighting the fire.

If the crew had hit the fire with the extinguishers immediately after discovering the pump did not work, the fire would have been slowed to the point that suppression could have been completed by the second due engine. Instead the crew became focused on fixing the engine, and consequently a car was burnt beyond repair. The nice thing about this example is that every department has a story that is very similar. You never hear the stories told at firefighting conventions, because there is no glory in the story, but every department has a story of a failure to maintain control leading to an undesirable result.

The key to maintaining control is to force yourself to take a step back from the situation, take a deep breath, and think to yourself, "What is really going on here? Am I seeing the big picture?"

Assess the problem in the time available

While the fire service operates under time pressures, rarely does a decision on a scene have to be made within the 30 seconds it takes to step back, gather good information, and make a good decision. The first question in the analysis is, "When does this decision have to be made?" The second question is, "What information do I need to make a good decision?" Very often we will make a snap decision, on limited information, when we can make an informed decision on good information, if we take the time to figure out where we are and what we need to do.

After assessing the time pressures, a good leader will shed the tasks for which the leader's attention is not required. Shedding can be accomplished by delegating the tasks to another person, or simply putting the tasks on hold, and working on the most critical problem.

For example the command officer can delegate handling radio traffic with dispatch and other divisions so the officer can concentrate on the problem area. The radio traffic does not go unanswered, and the fire scene does not grind to a halt because the command officer's attention is diverted elsewhere.

Another way of dealing with time-critical decisions is to preplan the response before the time critical event happens. The National Fire Protection Organization, in NFPA 1500 attempts to administratively regulate such a preplan when it imposes the two-in, two-out rule. Rapid intervention teams are such a preplan. If an interior firefighter goes down, the time-critical decision of who goes in to affect a rescue is already made, and the rapid intervention team is already equipped and ready to go. This example is an elementary one, but the pre-planning concept can be extended to all type of incidents.

NASA Space Division has taken preplanning to an extreme. Every contingency is planned and anticipated. If the relay and junction 3521A fails, astronauts can rest assured there is a procedure for bypassing the relay. NASA has contingency plan upon contingency plan. Every conceivable combination of failures is assessed and mitigation procedures established.

The fire service operates under a completely different set of variables, and variables that cannot always be anticipated. However, we do many things over and over and over again. If something goes wrong and things invariably do, having a contingency plan in place will save critical time in pressure situations.

Gather information from all sources

It almost goes without saying, that gathering relevant information from all sources aids the firefighter in making the best decision. How many times, in an after incident review, have we heard the statement, "If I'd have known that, I would have done something else"?

The strategy for gathering information is very simple. Start with the person who knows the most, and work your way to the person who knows the least. If you want to know about fire conditions inside the building, ask the person inside the building. If you want to know about structural stability on Division C, ask the person in charge of Division C.

Be aware of bias. There are times when someone may want to make an interior entry to save a trapped person, when such an entry is not advisable. The desire to make entry may result in incomplete reporting of actual fire conditions. A good command officer will know the people and be able to assess those biases.

Finally, every party reporting information needs to facilitate good communication. We are human beings and we are stuck with the inefficient and biased system of communication predisposed in human beings. We use words which may have several meanings, filtered by cultural biases, and personal experience. Information gathering should be done with a common language using the communication skills outlined elsewhere in this book.

Monitor the results

The best fire chief I have ever known has a rule. He says, "If you are doing something for a minute and it has not made a bit of difference, do something else because what you are doing is not working." If an action is taken, monitor the results of that action. If what you are doing works, keep doing it. If it doesn't work, stop.

CRM principles stress that all resources must be considered before a quality decision can be made. Therefore the quality of the decision is directly related to the amount and the accuracy of the information gathered pertaining to the crew, the fire, and the environment.

If a firefighter suspects that a loss of situational awareness has occurred, the firefighter should revert immediately to the basics:

- Maintain control—fight the fire

- Assess the problem in the time available

- Gather information from all sources

- Monitor the results—alter the plan as required

Use of this process will aid in anticipating accidents and injuries and prevent them.

Remember the three key words, *fire—plan—people,* and the questions that go with those words.

1. What is the fire doing?

2. How is the plan working?

3. How are the people doing?

Ask these questions early and often.

MEMORY

Human beings have two types of memory. Short-term or working memory retains small amounts of information while we are using other thought processes. The capacity of this type of memory is very limited and the retention of information in this type of memory is short term. We have items in our working memory for the time it takes to be acted upon, and then the information is either transferred to long-term memory or forgotten. Examples of types of information stored in short-term memory include looking up a telephone number or knowing what color the stop light is in front of your car while you are driving. Working memory is limited. We can store up to seven plus or minus two pieces of information in working memory.[16] The reason telephone numbers are seven-digits long is that is the average number of digits an average person can store in working memory.

Long-term memory is the storage center for information that may need to be retrieved at a later date, but is not to be acted upon immediately. The accuracy of long-term memory access depends upon frequency, recency, and relevance.[17] If an item in memory is recalled a number of times, it is easy to recall. If an event happened recently, it is easier to recall than years ago. If the event is relevant to someone's life, it is easier to recall. It is much easier to recall your birthday than the date when you got your braces off.

Effects of stress on memory

As an individual becomes more stressed, that individual's focus shifts from a group focus to an individual focus.[18] Additionally, when a person is placed under stress, that person's working memory diminishes. When a firefighter is beginning to panic because the firefighter is trapped in a building, that firefighter's working memory may not be capable of remembering the directions given to him outside on how to get out of the building.[19] On the other hand, a person's long-term memory may be enhanced by the stress.[20]

Given this knowledge about the functioning of the human animal, firefighters should be aware of the implications for training. We should factor the natural human response into the system.

When a person gets trapped in a building, we cannot expect that person to reason a way out of the building. We should minimize the working memory load.[21] When giving someone directions on how to get out of the building, we should not give the directions in one bulk transmission, but we should give three or four bits of information, and then allow that information to be digested before giving more information.

We should also over-train the individual. When a person gets in a stressful situation, the person normally reverts to prior over-learned behaviors. This phenomenon is a direct result of a person's working memory and reasoning processes breaking down when trying to think a way out of the stressful situation. A person cannot rely on reasoning abilities to think a way out of a problem.

We cannot rely on cognition to allow people to work through a stressful situation. We should train the processes until they become automatic, and embedded in long-term memory. When we do that, our operations become safer and more efficient, and our people are more prepared to react in a stressful situation.

STRATEGIES FOR INCREASING MEMORY

Some strategies for making working memory more efficient include using chunking, providing visual echoes, and over-learning.

Chunking

Chunking means putting the information in a relational group. The information, which is rationally related, constitutes one piece of information rather than several. Consider the information in the following chart. In one column, the words are unrelated and have no common theme or meaning. In the other chart, the words are all related and constitute a linked item. Consequently the information is remembered as one item rather than several.[22]

We can, through training, create a rationally related set of commands. A procedure, which is over-learned and memorized, becomes a song or chant like the song, "London Bridge is falling down." Much information can be transferred and remembered with a single related phrase (see Fig. 4–1).

Though	London
Minus	Bridge
Ramble	is
Celsius	falling
Fickle	down
Dogma	my
Tree	fair
Novel	lady

Fig. 4–1 Memory Chunking in Practice

Visual echoes

When possible, follow the command with a redundant visual echo. Either provide the command in writing or in a graphic, in addition to verbally. The redundancy and the alternate method provide additional connections in the brain, which aid in memory.

Over-learning

As discussed previously, an overly stressed person reverts to prior over-learned behaviors. How often on a fire scene have you found the stressed command officer on the end of a hose where the behavior was over-learned, instead of sitting in the command unit where the command officer should be sitting? That reversion to prior over-learned behaviors is a cue to the follower that the command officer needs to take a step back and figure out what is going on.

It is also a tool for training officers. If a person drills and drills and drills on a topic until it becomes automatic, when the chips are down that person reverts to the prior, over-learned behavior and the response becomes automatic, whether it is going to a left-handed search immediately upon entry or finding a charged hoseline in the dark and following it out of a room. We can exploit this trait in the human being to make sure when the unexpected happens, our people are ready to react. This is supported by the *training-in-context* theory, where you train as closely as possible to the way you play.

In one example a department received new Personal Account-ability Safety System (PASS) alarms and went though an extensive training process that almost mirrored the training-in-context theory. (This is an alarm that automatically goes off when the firefighter does not move for a short period of time, and it is used to locate fallen firefighters.) The firefighters were put in situations in the class that mimicked an actual structure fire. They bunkered up, slid on their SCBAs, turned on their PASS alarms (that they had just learned about), and entered the structure.

As everyone knows, PASS alarms on new people will drive them nuts because the alarms will go off constantly (especially the first models that we used). Therefore one step was added for benefit of sanity in training. The PASS alarms were turned on and then turned off immediately, prior to entering the structure. The trainers mistakenly thought that this behavior would not carry over to the fire scene. Unfortunately it did. Firefighters under stress were seen at actual incidents sliding on their SCBAs, turning on their PASS alarms, turning them off, and then entering the structure—prior over-learned behavior.

We find that if we encourage our people to over-learn a skill, they will make fewer mistakes than those with whom the skill is not over-learned. Over-learning, over-trained individuals made 65% fewer errors than a control group when retested eight weeks later.[23] People retain their skills and perform better over time if the skills are over-learned.

We also find an additional benefit. Highly practiced tasks can be task shared with other tasks with little interference in performance.[24] In other words if the task is over-learned, a person can cognitively perform other tasks while completing the task at hand, making the firefighter more efficient and more accurate.

CONCLUSION

CRM principles stress that all resources must be considered before a quality decision can be made. Therefore the qualities of the decisions are directly related to the amount and the accuracy of the information gathered pertaining to the crew, the fire, and the environment.

If a firefighter suspects that a loss of situational awareness has occurred, the firefighter should revert immediately to the basics.

1. Maintain control— fight the fire

2. Assess the problem in the time available

3. Gather information from all sources

4. Assess all the options—choose the best

5. Monitor the results—alter the plan as required.[25]

Use of this process will aid in anticipating accidents and injuries and prevent them.

REFERENCES

[1] Chappell, S. *Managing Situational Awareness on the Flight Deck, or The Next Best Thing to a Crystal Ball.* From Kent, T., *Flight Discipline,* New York: McGraw-Hill, 1998.

[2] This information is taken in large part and paraphrased extensively from the Crew Resource Management (CRM) developed by Transport Canada, System Safety, Western Region, this information is copyrighted.

[3] *Wildland Fires Safety Awareness Study: Phase III,* "Implementing Cultural Changes for Safety," Tri-Data Corporation, 1999: 5–62.

[4] *Ibid.*

[5] Schendel, J. D., and J. D. Hagman, "On sustaining procedural skills over a prolonged retention interval." *Journal of Applied Psychology,* 67, 1982: 605–610,

[6] Druckman, D., and J. Swets, *Enhancing Human Performance: Issues, Theories, and Techniques.* Washington, D.C.: National Academy Press. (1988).

[7] Orasanu, J. M, *Shared mental models and crew decision making,* Princeton, N. J.: Cognitive Science Laboratory, 1990; Cannon-Bowers, J. A., E. Salas, and S. Converse, *Shared mental models in expert team decision making,* 1993; Castellan, N. J., ed., *Current issues in individual and group decision making,* Hillsdale, N. J.: Erlbaum.

[8] Serfaty, D., E. E. Entin, and J. J. Johnston, "Team Coordination Training," in Cannon-Bowers, J. and J. Salas, eds., Making Decisions under Stress, American Psychological Association, 1999.

[9] *Fireline Handbook,* NWCG Handbook 3, PMS 410-1 NFES 0065, (January 1998).

[10] Driskell, J. and J. Johnston, "Stress Exposure Training," in Cannon-Bowers, J, and J. Salas, (eds.) *Making Decisions Under Stress,* American Psychological Association, 1999: 191, 212.

[11] *Thirtymile Fire Investigation,* Accident Investigation Factual Report and Management Report, Chewuch River Canyon, Winthrop, Washington, July 10, 2001, Forest Service, U.S. Department of Agriculture, September 26, 2001, amended October 16, 2001.

12. Ibid.

13. From Transport Canada, *Ibid*. See also Bush, D. *The Effects of Stress on a Pilot's Situational Awareness*, Proceeding of the Tenth International Symposium on Aviation Psychology, Columbus: Ohio State University, 1999: 1319.

14. *Ibid*.

15. *Ibid*.

16. Miller, G.A. "The magical number seven plus or minus two: Some limits on our capacity for processing information." *Psychological Review*, 63, 1956: 81–97.

17. Wickens, C., S. Gordon, and Y. Liu, *An Introduction to Human Factors Engineering*, 1998: 159, 162.

19. Wickens, C., S. Gordon, and Y. Liu, *An Introduction to Human Factors Engineering*, 1998: 385.

20. *Ibid*.

21. Loftus, G.R., V.J. Dark, and D. Williams, "Short-term memory factors in ground controller/pilot communication." *Human Factors*, 21, 1979: 169–181.

22. Wickens, *Ibid*.

23. Schendel, *Ibid*.

24. Driskell, *Ibid*.

25. Transport Canada, *Ibid*.

5

COMMUNICATIONS

"Communications problems are considered the most common operational snag in the majority of departments, effecting the firefighters' ability to start, coordinate and complete effective operations."

—CHIEF ALAN BRUNACINI[1]

IN A STUDY of FAA data concerning aircraft incidents, more than 70% of the reports contained evidence of error in the transfer of information. One of the most common communication problems was failure to initiate the information transfer process—37%. In most cases the needed information almost always existed, but it was not made available to those who needed it. Another 37% was inaccurate, incomplete, ambiguous, or garbled messages. Other problems included failure to transmit the message at the appropriate time, message was either not received or was misunderstood, and a small portion, 3%, was classified as communications equipment failure.[2]

Human factors issues related to interpersonal communication have been implicated in approximately 70% to 80% of all [commercial aviation] accidents during the past 20 years.[3]

As emergency workers, we depend upon a system of communications. By system I do not mean just the actual hands-on systems like radios or cell phones, I also mean the way we communicate and what is expected of us. As a group, we need to identify what information needs to go to the firefighter and what information needs to go to the chief. We should have systems that deal with how we transfer the information, what information needs to be transferred, and when this transfer of information should take place. CRM has a workable system of communications that includes inquiry, advocacy, listening, feedback, and conflict resolution. These are the basis of all cockpit communications between commercial airline pilots and can be easily adapted to the fire service. Keep in mind that all the technologically advanced equipment in the world will not cure a poor human system of communications. A study of aircraft accidents found

> *The primary conclusion drawn from the study was that most problems and errors were induced by breakdowns in crew coordination rather than by deficits in technical knowledge and skills. For example, many errors occurred when individuals performing a task were interrupted by demands from other crew members or were overloaded with a variety of tasks requiring immediate action. In other cases poor leadership was evident and resulted in a failure to exchange critical information in a timely manner.[4]*

TYPES OF COMMUNICATIONS

We depend upon communications every day. In the absence of other people, we still communicate, often talking to animals, trees, inanimate objects—such as stoplights, and, as a last resort, ourselves. We can seldom go very long without communicating some type of information. This information can be verbal, non-verbal, written, or symbolic, and is most commonly a combination.

Verbal communications

Verbal information is passed face-to-face, on the phone, and over the radio. This is what people usually think of as communicating. Taken at face value it can mean one thing, but look deeper and you may find an entirely different meaning.

> *It is not what you say, it is how you say it....Two dialogues really take place in every conversation—one uses words, the other tone of voice....When you ask someone, 'How are you?' and get the reply, 'Fine,' you are not usually relying on the word 'fine' to tell you how they feel. Instead, you let their tone tell you whether they really are fine, or whether they are depressed, anxious, excited, or feeling any of a dozen other emotions.*[5]

Besides the words, we tend to listen to the tone of the communicator's voice, the volume, the speed at which the communicator delivers the message, and word selection. All of these help us form our impression about the person, the message, and how to respond.

If we are en route to a structure fire, we can hear the seriousness of the call in the dispatcher's voice or the voice of the first unit on-scene. Experienced personnel usually deal with fires very calmly and professionally. Their selection of words is very familiar to the fire family; we've heard them many times before. The tone and speed of the message conveys the severity of the incident.

Think of your last structure fire. If it was a typical single-family structure with no one trapped, it was probably handled very deliberately because we have responded to those kinds of calls hundreds of times. But add more and more stress to the incident, like fully involved, occupants trapped, *children trapped*, and even those professional voices tend to get a little higher and a little faster, and maybe they lose some of that polished jargon that we have become so familiar with. Unusual incidents and situations tend to *lock-up* our mental processes. We find it hard to find the right words and we sometimes blabber over the radio without spitting out our intended message.

I remember one hazardous materials incident where I was attempting to convey the proper spelling of a 30-character chemical name and I was running out of time. I was spelling the name phonetically, and when I got to *g* I could not remember that phonetically it is pronounced *golf*. I used all the brainpower I had in reserve and came up with a new phonetic alphabet that now includes <u>God</u>.

Another problem is that we become so casual—or uninvolved—in our defined speech patterns that we speak without thinking.

> ...*[There is] evidence in the data suggesting that the standardization of...language can lead to an unwanted side-effect on the mechanisms of language production: The fixed phraseology may be automatically retrieved from the lexicon, so that it would not go through the conceptual stages of language production.*[6]

An example of this phenomenon may be found when responding to a call. The officer and driver are communicating about hazards, etc. while en route. The usual practice is for the officer to look to the right and verbalize if it is clear, "Clear on the Right," or if there are hazards, "Brake Right." These are the language and statements (the lexicon) that are standardized, defined, trained, and approved by the department in which they operate.

While responding to a potentially stressful call, an officer might say, "Clear Right...brake! brake! brake!" The verbal communication has become standard, a *stock phrase*, and does not have the intended cognitive processes that should accompany it. In other words, we open our mouths before we think. This is made easier because we have standardized our language to a point where we do not have to think first (visualize what we see at the intersection). Then we formulate a phrase to communicate (clear right or brake right), then communicate our thought. Be careful when communicating these stock phrases. Make sure that you are thinking before speaking.

Experienced fire personnel, especially leaders, realize that increased volume, increased speed, and a raised tone in their communications do not really instill a lot of confidence in the troops. Leaders have become experts at disguising the stress in their vocal characteristics. They are the capable and confident leaders whom people look to for support and encouragement when the incident seems out of control.

Written communications

The written word has become more common and has replaced a lot of verbal communications in the modernized world. Now we can e-mail someone across the world easier than we can give that person a phone call. Drawbacks do exist in this form of communication. Many times people read messages only to become enraged because they *perceived* the tone or intentions of the sender incorrectly.

In the absence of other forms of communication, we use our perceptions to interpret what was meant. The choice of writing style could affect the message. Does a fact-based memo sound too cold to send to a friend or associate? Does a letter requesting information make it sound like the reader is required to do this? Does a friendly, rambling e-mail make the sender look unprofessional to an associate? A funny example is found in this excerpt:

> ...*I said [in my memo] that even though we had accomplished a lot the past year we couldn't stand pat during the coming year. A day after the memo was handed out, a woman asked to see me and then, after breaking down into tears, asked what did I have against a coworker of hers, a woman whose first name was Patricia. It seems Patricia was herself very upset and crying in the ladies' room because the both of them couldn't understand for the life of them why I couldn't "stand pat."* [7]

Non-verbal communications

Non-verbal communications are very important for sending messages too. How many times today have you received a non-verbal communication—one without any words? Think about the last time you cut someone off while driving your car on a busy street; did you need to hear that person's words to receive the message? There are literally hundreds of different body movements, gestures, postures, facial expressions, gazes, pauses, etc., that can clue us in to the real message behind the words a sender uses. A nervous glance down at the shoes by the sender may indicate that person is not being altogether truthful about what is being said. A person's perfect posture with hands at the sides, and straightforward eye contact may indicate confidence in a message. Invasion of a speaker's personal space may mean a date for tonight. All these intricacies of human interaction can speak volumes if they are used.

A paramedic who is normally very skilled and professional may fumble through the medical kit looking for the right drug to administer, or miss important information. While this behavior is uncharacteristic for this provider, and this person continues to say the *right things*, the actions may be a clue as to the state of mind at that time. It may be a personal problem that has been brought to work; it could be this person is struggling with drug abuse, or it could be that it is just an *off day*. Whatever the case, the immediate presentation of non-verbal communications screams out for assistance. You must apply the non-verbal cues to the verbal cues to come out with the complete meaning of a message.

Symbolic communications

Symbolic communication consists of our appearance and what we keep near us in the way of items—our knickknacks. If we meet a person on the street who is wearing an expensive suit and has perfectly styled hair and spotless shoes, we might see that person as a business professional. But at home, we see the same person in comfortable, weekend clothes with holes and stains. At first glance, we might think that person is less professional. That same person's office is decorated with college memorabilia, community awards, and gifts from all over the world. This might indicate a well-educated, community-minded professional with friends and contacts in many different places. At home this person's garage is a mess, the lawn is half dead, and this person doesn't even know the neighbors' names. This may indicate that a person does not spend much time there or that work is more important.

There may be other conclusions that you draw about the discrepancies between the *business* environment and the *home* environment, but that is only part of the picture. We think we can tell a lot initially about who a person is by appearance and surroundings. Unfortunately, symbolic communications can be easily staged, and therefore mis-interpreted by others. In the above example, it is hard to tell which is correct, the home or the business symbology. It could be that either one is the *real* person.

Symbolic communications are but a piece of the human puzzle and must be used with other pieces of communications to make an accurate assessment. Other forms of symbolic communications can be religious symbols, sports memorabilia, clutter, cleanliness, hygiene choices (haircuts, fingernails, etc.), jewelry, glasses/contacts, and hundreds of other things that people choose to display to others.

In the fire service, symbolic communications run deep. Firefighters are proud of their affiliations and it is not uncommon to see department or station affiliations displayed on hats, T-shirts, or jackets worn by firefighters, even while on non-department business. This shows pride in what they do, who they are, and maybe subliminally asks the observer for respect. Conversely, wearing a T-shirt that says "Harley Fire Department—Find 'em hot, Leave 'em wet," sends a completely different message to the observer—one that does *not* exactly shout pride and professionalism.

What if Computers Operated Like People?

Think of it. What if computers operated like people? What if you had a computer that changed the way it reacted to the same commands given to it at different times of the day? Its understanding of what you want done, whether it will do it, and to what degree is affected by how fragmented its disk drive is, how full its memory is, what kind of power fluctuations it has had to deal with, and the workload you have put on it earlier. After taking all of that into account, it may or may not perform the command. It may or may not do it *when* you want and it may or may not do it *correctly*. Also, what worked well for communicating commands to this version of technology, probably will not be the most effective on a newer version.

I will bet you would not keep that computer around for very long under those conditions (I know I didn't keep mine). Communicating with people is exactly like that. So many things affect their input and output that it is amazing we accomplish much at all. Luckily much like a computer though, people are somewhat predictable in the way they will communicate—although what they communicate is sometimes still a mystery. A lot of research emphasis has been placed on the communications side of human interaction. It is still a little fuzzy sometimes, but there are some general guidelines and practices that will make communicating with your fellow computers—er— people, a little more reliable and prevent them from *locking-up* during those stressful events on an emergency incident.

WHY ARE YOU COMMUNICATING?

Of course the main reason we communicate is that we want others to share the exact meaning for a thought, feeling or idea. We attempt to share this meaning through the exchange of verbal, written, non-verbal, and symbolic communications. As you well know, these systems of communications each have its own advantages and drawbacks.

I just saw a comedian perform on one of those talent shows on TV. He had a great bit about his dad trying to tell him where to find a part in the garage. Dad's hands were full and he was trying to get the son to grab the *thing-a-ma-jig*. The proceeding head gestures, incomplete and ambiguous verbal instructions, and obvious, non-verbal, irritation were hilarious. How many times have you received an important instruction through the general wave of a hand and an instruction like, "Get that thing. Not that one, the other thing. No! No! No! More left. More up. Left. No! Up Up Up!" This comedian had definitely hit the communications problem on the head.

In emergency situations we are dealing with time pressures, important information, and varying levels of experience and backgrounds. In addition we have some almost insurmountable obstacles, such as talking through an SCBA, high noise levels, unknowns, and many other problematic issues. Anyone who has been in the fire service for very long has probably received one of the, *There's a fire under the stairs! I need more hose!* communications from someone. That person is standing in full bunkers and SCBA, in a billowing smoke-filled doorway, complete with expletives, hand-signals, and an unstable look in the eyes.

What we probably heard under stress was, "there's a choir under the squares! I need more clothes!" Something that is very important to communicate may get rushed, abbreviated, and/or misunderstood. All these factors make the fireground a very dangerous place for firefighters—and a very difficult communication minefield.

Because of all these factors, we need to design and practice a system of communications. This provides for a common framework of communications that becomes the standard within the context of all fire department communications—on-scene or off.

It's a fact—the way we communicate

After examining tapes of actual cockpit voice recordings prior to and during emergency situations that were a matter of life and death, researchers found that

> ...communication is always coherent under normal low and normal high workload conditions. In contrast, ...danger situations contain instances in which the communicational dynamics are disrupted: questions are not answered; the crew members work on different communicative tasks and interrupt each other. These results suggest that the ability to produce coherent dialogic structures is impaired under the influence of time pressure and danger. That is, under the danger conditions it seems to become more difficult to include the utterances of the communicative partner into one's own planning processes.[8]

It is human nature. We cannot help running into problems when we try to communicate our thoughts, feelings, and ideas to another human being. It takes a lot of work and self-assessment to change the way we are programmed by nature and by society, as well as, our upbringing and personal values. While day-to-day life does not really necessitate that we change the way we communicate, the emergency incident begs us to be aware of our shortcomings and change the way we communicate to save our life—and our brothers' and sisters' lives. All of the following items describe us as a whole—naturally. Read these Eight Realities to Communications in Humans and try to honestly evaluate yourself on and off the emergency incident. Have you ever done these things?

Eight Realities to Communications in Humans

1. We tend to protect, maintain, and enhance ourselves when we communicate.
2. We defend against looking ignorant or foolish for fear of ridicule.
3. We wish to maintain consistency; we tend to support our opinion even when we suspect that we may not be totally correct.
4. We wish to feel valued, worthwhile, belonging and meaningful. This means that we must be acknowledged with respect and trust.
5. The reality of the situation is second to our perception of the situation—and our mind set may be very difficult to change.
6. We behave according to our perceptions; we may not be aware of the level of risk.
7. Emotions take first place—feelings are the facts.
8. Commitment comes from self-determination; people have their own motivations.

—TRANSPORT CANADA CREW
RESOURCE MANAGEMENT HANDBOOK

ORDERS ARE MADE TO BE FOLLOWED

Studies show that constant one-way communication, usually orders or the like, does not improve team performance, especially during times of stress. One-way communication tends to indicate one or more team problems. It could be that the leader has established himself as sole decision-maker and communicator and will not tolerate feedback from other members. Or it could be that the leader is uncomfortable with members' abilities, the team members themselves, and/or their knowledge of the situation. Or it could indicate that the subordinate is overloaded or is trying to *figure out* the situation before he speaks up.

It is not that the subordinate disagrees with the leader; it may only be that his perception of reality and the best course of action do not match those of the leader. This silence should be a sign that your team is not performing as well as should be expected and may be wasting time trying to figure out the leader's decisions instead of just inquiring about the situation and the decision. I would venture a guess that at least 95% of the decisions that are made on the fireground are made too quickly—without enough information-gathering, discussion, and review. The other 5% of the decisions on the fireground are appropriately snap-decisions made because the situation dictates immediate decisions and orders without the luxury of time.

We tend to use the emergency scene as an excuse for all types of irrational or inappropriate behaviors. We say that the scene requires immediate actions (most of the time we have at least 10 seconds to step back and look at what we are about to do). We say that the scene requires orders—not discussions (if it might kill me, I would want to know what is going on and how to make it out alive). We say that the scene is no place for training (what better place to really teach someone the right way, and maybe the wrong way, than an actual call). We must use 95% of the calls to build a strong, functional, knowledgeable, and trustworthy team for that 5% of the calls where we must rely on each other for immediate decisions and actions without the luxury of time.

Effective communications between crew members ought to be maximally explicit and direct.[9] This means that you should call a duck, a duck, because being politically correct and calling it a waterfowl leaves the exact definition up to the receiver as to whether you are talking about a duck, goose, or swan. A team should be comfortable enough with its members to accept this direct and explicit communication style in times of stress—such as on an emergency incident.

> *"...the capability to adjust readily to unexpected events might be hindered by the existence of competing standardized tasks. When faced with competing communicative demands we do not respond to the unexpected task but prefer the one that belongs to the script. To depart from the script requires higher processing capacity, which is particularly difficult in situations of already high cognitive workload."*[10]

GUILTY AS CHARGED— ASSUMPTIONS THAT COULD KILL YOU...

A couple that had been married for a while was having some serious relationship problems. They decided they should see a marriage counselor.

> *She started out by telling the counselor, "I really found out what I was in for on the first morning of our honeymoon. He made me toast. We had a whole loaf of bread and he made me two heels! I should have called it quits right there."*

> *All he could say was, "I gave you the heel because it is my favorite part and I wanted you to have it."*

In the context of a staff meeting, you may not even notice some of these assumptions in our communications. You may observe some of the problems, but they are usually correctable in the "slow-paced" administrative setting. In contrast, the emergency incident sometimes only gives you one shot at getting your message across—correctly.

Therefore, knowing where those important communications will break down, will prepare you to deal with them prior to delivering your message. You can plan and practice, in training and in the fire house, what you are going to communicate, how you will communicate it, and what to expect from the receiver of the communication before you make a high-risk communication on the scene.

What I said is what you heard

Wrong! Take this sentence for example. You are reading it, or are you? Have you read every word? Do you understand all of the vocabulary? Do I have an annoying habit that prevents you from concentrating; i.e., misspellings, improper grammar, repeated use of certain words, or rambling paragraphs with no points (guilty)? So even what you think would be the easiest form of communication, really has a lot of barriers that could affect the transfer of a message. With verbal communications we do not have the luxury of re-reading what someone has said, and if we do not question (inquiry) what they have said, their exact meanings, etc., we are left confused and ineffective.

I only communicate when I want to

Wrong! You communicate continuously through non-verbal and symbolic cues. You cannot stop communicating. If you just stand there, do not move, and do not say a word, you are still communicating something (quite possibly, fear of being eaten by a large dinosaur-like creature). Silence is still a form of communication, sometimes, even more powerful than words. Fire instructors know this routine all too well. They can watch their students and see the level of understanding or misunderstanding on their faces. Realize that on an emergency incident, some of the most powerful communication is the kind that you do not hear.

My message depends only on the words I use

Wrong! Some words have multiple definitions and multiple perceptions. Take for instance part of the radio transcript on a structure fire in a small mid-western community. In a size-up of the incident the IC said, "…We got a large storage building. It appears to be…it's fully involved in fire…." Define in your own mind, *large*. My definition will be different from yours, I am sure. Is large a 100' x 200' building? How about 40' x 60'? Or, maybe, a large shed is 12' x 20'. Not that I need the same information at this moment, but when our lives depend on communication, this is a problem to address and correct.

Have you heard the one about multiple meanings for words? Three members of a cowboy separatist group were captured by the Omega Force and sentenced to die by firing squad. Clem was the first to appear. The Commander asked if he had any last words, and he thought for a minute and he yelled, "Flood!" All the firing squad ran for high ground and Clem was able to escape. Bob was the next cowboy to appear. The Commander asked if he had any last words, and he thought for a minute and yelled, "Tornado!" All the firing squad ran for the shelter, and Bob was able to escape. Bubba was the third cowboy. The Commander asked if he had any last words. He thought for a minute and then yelled, "Fire!"

Another good fire example happens when the IC sends a team in to check the first floor of a high-rise structure. Firefighters are always asking their partners, "Is the first floor the ground floor or is it the first floor above ground level—or is that the second floor." Double meanings, slang, and complexities in language make misunderstandings commonplace; address these immediately to prevent bad things from happening.

I do not need to repeat an order; they got it the first time

Wrong! Think back to how many times an address is repeated over your radio en route to an emergency. In a radio transcript from a structure fire, a street address was repeated 12 times. Still the street was mistakenly called something else eight times. Another address was stated six times and was still mistaken four times. If you do not think you need to repeat something as simple as a street address, how about a complex set of directions?

Try this exercise. Sit someone in front of a class and read about 15 items into his ear for a minute. Then ask the student to write down all he remembers. If the student is normal (normal firefighter, I mean) he will remember about 7–12 items. Now put that firefighter under stress and he will only remember about 3–6 items. Remember that! On an emergency incident, you either need to *over-train* on the SOPs, change the number of verbal orders given, or change the method for giving extended orders—more on this technique in the decision-making chapter.

It is not my problem if he cannot understand directions

Wrong! It is not always the receiver's fault when he does not understand the message. There is an old instructor's saying that goes, "If the audience is asleep, wake up the speaker." How many times have you personally been in a class (or conversation) that you just could not get interested in and really were not listening? Maybe the verbal, non-verbal, and symbolic communication styles do not match. For instance, what if the IC on a large fire looks very stressed, disorganized, and a little frazzled? Maybe he keeps running around the fire in shorts and a T-shirt and will not stay put at the command post. No matter what information he has, you probably will always be a little concerned because the non-verbal and symbolic communications don't match the verbal ones or the requirements of the situation.

If we just had better radios, we would not have these problems

Wrong! Most communication problems involve *liveware* (humans) and not hardware (radios). Look through the radio transcripts and count the number of questions that get no answers, questions answered with questions, statements made to nobody, and repetition of addresses. All of these items would not be solved if we had live truck-to-truck satellite video hookups (well, maybe with video hookups). Remember in the reference earlier in this chapter? Only 3% of communications problems were related to technical problems.

I AM AWARE OF THE PROBLEMS— WHAT NOW?

...crew performance was more closely associated with the quality of crew communication than with the technical proficiency of individual pilots or increased physiological arousal as a result of higher environmental workload. No differences were found between the severity of the errors made by effective and ineffective crews, rather, it was the ability of the effective crews to communicate that kept their errors from snowballing into undesirable outcomes.[11]

Good communicators are not born—they are made. It is a group effort and everyone has to be on board for it to work effectively. Below are some items to consider any time you are communicating with anyone over any type of medium—radios, computers, face-to-face, etc. *...[researchers] have found that human factors training produces measurable improvements in communication skills.*[12]

The number one item to remember is that it is NOT MUTINY, it is TEAMWORK. There have been concerns both in the aviation community and the fire community that teaching a subordinate to follow the CRM System is disrespectful. A quote from Bob Thaves, the cartoonist of *Frank and Ernest*, sums up the philosophy well. "Question authority, but raise your hand first."

Interactions between fire personnel must always be respectful regardless of rank. Assertiveness is required to be part of the team, but it must always be respectful and appropriate. The steps of inquiry, advocacy, listening, feedback, and conflict resolution are vital to the team's safety and the efficiency of the operation. Also vital to safety and teamwork is the process of monitoring and challenging your teams' decisions and actions.

Another tongue-in-cheek quote might fit also, "He may not always be right, but he is always the chief." Like it or not, fire scenes boil down to one person who is in charge. That person has the responsibility and the authority for that incident. Fire scenes are not a democratic operation. ICs must gather information and make decisions—that is their job. Remember that you can give feedback to ICs, but ultimately it is their choice to make the final decisions. If the decision is wrong, it is the IC's responsibility. Of course if it boils down to life and death, a firefighter should always have the right to decline an assignment. *What?*

At a CRM Symposium, I specifically asked researcher Lynne Martin, then of the University of Scotland at Aberdeen, what I should do if a subordinate did not believe an assignment that I gave was safe. What if I am completely comfortable about an assignment and its safe and effective outcome; but a less experienced firefighter does not feel safe with my decision. What should I do? After about an hour, we both agreed upon a few important items:

1. Firefighting is a very dangerous job that requires extreme amounts of teamwork, courage, and intuition for a successful outcome.

2. Because of the teamwork required, people on the team must be on the same page. If I wanted to go interior on a suppression assignment, but the probationary member was not comfortable, I might be able to pressure or order the member into following me. The problem is that as a team, we both have duties and responsibilities that require our full attention. If one of us is worried about our backside, what kind of team member are we going to make?

3. Therefore, we *must* take the time to make sure that all team members are comfortable with the decisions/actions, either through better training, better pre-briefings, more complete discussions, or alterations of operations to make the best possible use of the team.[13]

THE COMMUNICATION SYSTEM

Now, we know that communications are of utmost importance on the fireground, and we know there are a multitude of problems that can affect those communications. Here are the tools imbedded in an effective communications system that will increase the level of teamwork and understanding that you experience. There are basically three simple steps—inquiry, advocacy, and monitoring.

Inquiry

Numerous simulated scenarios have been observed in which one crew member had valuable information that was not communicated for one reason or another.[14] In the fire environment we gather information many different ways. We feel what the weather is doing. We feel the heat. We look to see if the roof is sagging. We look at the color of smoke. We hear what the fire is doing. We review building plans and construction techniques. In many of the same ways, we look to other firefighters to build our information base through communications.

I have a saying, "You don't know what you don't know." That simply means that if you have no idea what is happening, and you have no way of questioning it or changing it—that is very dangerous. Never be embarrassed to ask a question, no matter what level of training or what rank you have; what you don't know could kill you. Your pride will often be restored when a fellow firefighter reinforces information, corrects misinformation, or states, "I thought you knew," and respects you even more for asking that person for input.

To clarify an order or expected action is always a right of any firefighter at any level

If I do not understand what I am supposed to do, how am I supposed to do it? Again, inquiry is always respectful. Inquiries are usually questions because I do not really agree with, or understand, what we are doing with the training, knowledge, experience, etc. that

I currently have—it is like my asking, "What am I missing here, Cap'?" Many times an inquiry step is all that is required to clear up any misunderstandings and confusion.

Inquiry example

FIREFIGHTER: "Why are you adding dish soap to the hose stream?"

CAPTAIN: "It makes the water wetter and overhaul is quicker and easier."

FIREFIGHTER: "Wow!"[15]

End of discussion. Sometimes, inquiry does not explain why we are doing something, and the individual inquiring still feels the need to clarify an operation or make a change in operations. Then, the advocacy step must be used. In order for cockpit crewmembers to share a "mental model," or common understanding of the nature of events relevant to the safety and efficiency of the flight, communication is critical.[16]

Inquiry Skills

1. Be Respectful in your communications.

2. Ask for clarification.

3. Use questions.

4. Actively seek answers for things you do not know, even if they seem unimportant right now.

5. Verify your information and information transferred to you.

6. Be explicit and direct in your communications.

7. Admit any confusion or misunderstanding.

Advocacy

This is the step that makes everyone nervous. For years we have been told, or told someone else (usually someone under three-feet tall), "Don't talk back to me." Now we are changing the rules of the game and saying that everyone who disagrees with a decision or has reservations about their abilities to perform a requested action, has a duty to advocate their position. Again we must do this respectfully.

If I start out an advocacy statement with, "Boy that was stupid to send a truck…" How much of that message do you think got through to my intended receiver? Or what if I advocate my position and the IC replies with something like, "No wonder you're just a firefighter…" What type of interaction are we going to have now or later? The firefighter will probably not speak up just to see the IC's plan fall apart. Also, the IC will probably treat the firefighter like a troublemaker.

Unfortunately lack of communication is what leads to firefighters' getting injured or killed, and makes our operations less than professional and efficient. A skilled IC knows that he does not have all the information or the proper perspective on the incident all the time. Therefore, the IC should expect some feedback on decisions. In fact, in a discussion regarding advocacy, numerous command officers expressed personal concern that firefighters did not feel like they could or should advocate their positions or concerns. A NTSB review of air carrier accidents from 1978–1990, identified 302 specific human errors as factors in 37 separate accidents. Eighty-four percent were communications errors, specifically errors in monitoring and challenging skills. Seventy-five percent of these errors were influenced by operational and organizational conditions where external factors combined with basic human deficiencies resulted in an accident.

What does this mean? It means that 8 out of 10 subordinates would not tell the officer of an actual or perceived problem that could result in their own death, the death of their coworkers, and charges. It means that the culture and relationships that we nurture will affect the ability of subordinates to speak up when something is not going right. Teams need to remember this statistic and make sure everyone is able to inquire, advocate, and monitor what the team is doing. Leaders need to encourage feedback and ask for it if it is not coming freely. Firefighters are not expected to blindly follow their leaders into a known deathtrap, and leaders should not allow firefighters to remain silent while they are taking that trip.

Advocacy Example

CAPTAIN: "Pull another $1\frac{1}{2}$" to the front door. We're losing it."

FIREFIGHTER: "Why are we pulling another $1\frac{1}{2}$" instead of a $2\frac{1}{2}$"?" *(Inquiry)*

CAPTAIN: "Because we need more water quickly!"

FIREFIGHTER: "If we pulled the Blitz Line we'd have twice as much water just as quickly." *(Advocacy)*

CAPTAIN: "Excellent idea! Pull the $2\frac{1}{2}$"."

This is a successful communication cycle using inquiry and advocacy. "It's not mutiny; it's teamwork." The captain could have suffered a loss of situational awareness that prevented him from seeing the actual situation, selecting the right option, and making the right decision. The above communications cycle was not very controversial, but some can become quite heated. Remember to separate the people from the problem. "It's not who's right, it's what's right." Remember something that goes along with that statement; "It may not be wrong, it may be that it's just different."

A short story from Chief Ron Coleman illustrates this concept (forgive me, Chief, if I mess it up; it has been years since I heard this first-hand).

At a structure fire, a safety officer rushed up to the command vehicle where Chief Coleman was in charge. The safety officer stated that he would have to shut down the roof operation because the ladder the firefighters were using was set-up upside down. Chief Coleman calmly asked him if there were firefighters on the roof. The safety officer stated that there were. Chief Coleman said, "Good. I didn't tell them how to get on the roof—I just wanted firefighters on the roof. We'll deal with your issue in training."

So you can see in this example that the main mission was to get firefighters to the roof, and not how to get them to the roof. It would not have been my first choice, but it did the job—"Different, not wrong." You and your firefighters need to realize that there are a thousand different ways to do the multitude of items we are required to complete on the fireground. Some are dictated by regulations, some by standards, some by SOPs, some others by environment, and still others by necessity. Even those that are regulated may have options allowed when in the scope of a dynamic fire operation with variable factors.

Keep in mind that there are usually twice as many ways to do something as there are firefighters on scene—but all firefighters will usually "fight to the death" to convince you their idea is the best. Beware of those individuals who claim divine ability and knowledge to apply their trades through the use of *common sense*. This common sense is built by years of doing and seeing and learning, but may not always be based on the best track record or the best information. The short answer—common sense is not an excuse to be disrespectful, or a reason to stop following a strong inquiry/advocacy model of communicating information.

From another perspective yes, I know the IC has ultimate authority to make decisions, and as a firefighter, I should follow those even after I have advocated my position. What if I still think it is too dangerous? What should I do? I would probably assert my position again. If I still feel uncomfortable, I should state that in my advocacy statement. Feelings plus the facts usually have a place in the second advocacy statement. "I feel that if I get on the roof that it will probably collapse and I will be hurt or killed. Isn't there another way we can ventilate? Maybe horizontal positive pressure?"

If I still feel uncomfortable, the IC should make a change in the decision or take more time to explain it. Why? Because, let's face it, if you are supposed to perform a task that is integral and vital to the success of the operation, and you are so worried that you are not going to make it out alive, what kind of team member are you going to be? Will you be one that contributes to the team and the incident, or one that is so concerned about getting hurt or killed that you do the task halfway, possibly putting other firefighters at risk?

On a recent wildland fire, multiple crews turned down a very dangerous assignment—building fireline downhill to a fire. The leadership team was very obnoxious and ordered crews to perform the assignment. One after another turned it down, until a crew with a known culture of risk-taking accepted the assignment. Crews that had turned down the assignment had been re-assigned to areas of mop-up, or camp patrol (punishment for disobedience no doubt). The crew that had taken the assignment was rewarded with better assignments in the following days, more hours for their shift (crews like the overtime), and kudos in meetings.

Of course the assignment went as planned, but was it luck or not? The reality is that people operate from a base of what they know. The perceptions and levels of risk they are willing to accept guide their actions. Leaders should not force assignments on people who are uneasy about what they are about to do. Discussions should take place so the perceptions of both leaders and followers more closely match each other, which hopefully will bring them both closer to the actual reality of the situation.

Remember! Be assertive. If you were concerned enough to bring it up initially, stay with it. Get the information you need—your life may depend upon it. Be respectful, but forceful. I will not accept a suicide mission and I do not expect anyone else to either. But remember, a lot of experience and knowledge can be gained by being mentored by a more experienced firefighter.

Here is an excellent example from commercial aviation of what can happen when misunderstandings exist and inquiry/advocacy does not.

March 27, 1977, Tenerife, Canary Islands, Pan American Flight 1736, Boeing B-747-121, KLM Flight 4805, Boeing B-747-206B.

Both aircraft were diverted to Tenerife because of a bombing at Las Palmas Airport. After an extended delay, both planes were instructed to backtrack up the runway. The KLM plane reached its takeoff point, while the Pan Am plane was still on the runway. The Pan Am plane continued up the runway missing the taxiway turnout. There was heavy fog on the runway. The KLM plane began its takeoff without permission from the air traffic control tower, with the Pan Am plane still on the runway. The KLM plane had just lifted off the ground when it hit the Pan Am plane as it was taxiing. Both planes burst into flames. KLM had 234 passengers and 14 crew; Pan Am had 326 passengers and 9 crew; a total of 583 people were killed.

COCKPIT VOICE RECORDER TRANSCRIPTS (EDITED).
(PROBLEMS WITH COMMUNICATIONS ARE ADDED BY AUTHOR.)

Radio transmissions are broadcast between the aircraft and tower, or vice-versa, and are heard by other aircraft. Other communications within this transcript are intra-cockpit, heard only by crew members of that aircraft.

KLM TO TENERIFE TOWER: Uh, the KLM...four eight zero five is now ready for take-off...uh and we're waiting for our ATC clearance.

TENERIFE TOWER: KLM eight seven zero five, uh you are cleared to the Papa Beacon, climb to and maintain flight level nine zero right turn, after take-off proceed with heading zero four zero until intercepting the three two five radial from Las Palmas VOR.

KLM TO TENERIFE TOWER: Ah roger, sir, we're cleared to the Papa Beacon flight level nine zero, right turn out zero four zero until intercepting the three two five and we're now at take-off.

KLM CAPTAIN TO TENERIFE TOWER: We're going.

Problem: "We're going" isn't standard phraseology. It doesn't identify who is going, or where they are going.

TENERIFE TOWER: OK.

Problem: This affirmation is directed at an unidentified receiver. Radio communications must be explicit in the Who aspect and in the read back of critical information. Both are lacking in this transfer.

PAN AM FIRST OFFICER TO TENERIFE TOWER: No...eh.

Problem: This transmission is an interruption into a previous conversation; but, because the receiver of the last message was unknown, assumptions kick in.

TENERIFE TOWER: Stand by for take-off, I will call you.

Problem: Again, unidentified receivers.

PAN AM FIRST OFFICER TO TENERIFE TOWER: And we're still taxiing down the runway, the clipper one seven three six.

Problem: Equipment interference. PA radio transmission and Tenerife Tower communications caused a shrill noise in KLM cockpit—messages not heard by KLM crew.

TENERIFE TOWER: Roger alpha one seven three six [the Pan Am clipper] report when runway clear.

PAN AM FIRST OFFICER TO TENERIFE TOWER: OK, we'll report when we're clear.

TENERIFE TOWER: Thank you

PANAM, CAPTAIN: Let's get the hell out of here!

PANAM, FIRST OFFICER: Yeh, he's anxious isn't he.

PANAM, FLIGHT ENGINEER: Yeh, after he held us up for half an hour. Now he's in a rush.

KLM FLIGHT ENGINEER: Is he not clear then?

Problem: This is an inquiry statement—a question about the location of the other aircraft on the runway. This is a probable response by the flight engineer to actions being taken by the captain and first officer that indicate they did not receive the message that the Pan Am Clipper was still on the runway.

KLM CAPTAIN: *What do you say?*

Problem: A hint that maybe the captain is losing situational awareness. He may be overloaded with tasks related to takeoff in poor weather and is unable to communicate with other crew members appropriately—especially those messages that do not conform to the "script" of normal operations.

KLM UNKNOWN SPEAKER: *Yup.*

Problem: Unknown communication, sender/receiver/message.

KLM FLIGHT ENGINEER: *Is he not clear that Pan American?*

Problem: Second inquiry statement by the flight engineer about the location of the other aircraft on the runway. This question is a little more direct than his first.

KLM CAPTAIN: *Oh yes.* [emphatically]

Problem: The flight engineer either had a sense that things were not right or had understood the communications between the tower and the Pan Am Clipper; and, that the runway was not clear. Instead of a second inquiry statement, a more pointed advocacy statement might have been appropriate.

[KLM begins its takeoff roll]

[PanAm captain sees landing lights of KLM at approximately 700 meters]

PANAM, CAPTAIN: *There he is … look at him!* [Expletive deleted] *that* [Expletive deleted] *is coming! Get off! Get off! Get off!*

TRANSCRIPT ADAPTED FROM AVIATION SAFETY NETWORK
http://aviation-safety.net/cvr/cvr_kl4805.shtml

The lack of communication, understanding about the situation, and usage of outside resources is evident in this example. Do not assume that the information you missed is not, or will not be, important. Question, Question, Question. Question yourself, others, the situation, and the plan.

Advocacy Skills

1. Be respectful—regardless of rank.

2. State your opinions, or recommendations, about the information or your team's chosen course of action—use personal feelings if necessary.

3. Suggest solutions.

4. Recommend alternate actions.

5. Get an answer to your questions. Don't just throw your concerns out there and leave them unattended.

Monitoring

Monitoring is the process of keeping track of the effects of your actions. This is especially important to do after an inquiry or advocacy statement or discussion. If you make a mutual decision, and then nobody monitors the effect or the outcome, it is very likely that the outcome will not unfold as expected. The fire service is very good at initiating actions, but not very good at keeping track of the impact that those actions have. In reality this is feedback.

Many times a firefighting team will set goals or "triggers" in their discussions and planning. "If the fire gets to the other side of the road, we'll activate mutual aid." "After a fire has been impinging on the anti-gravity mechanisms in a building for 15 minutes, we should pull out." Good teams set up these points in advance because they know that one individual cannot monitor all that is required. But a team can keep track of some of these items, watch for them, and communicate them to the IC when they occur.

Listening

The failure to really listen to a message can be cited in many injury and fatality scenarios. How many times do people ask a question, and then not wait for the answer. Or how many times is an answer repeated. I hear you, but I am not listening is very common in fireground communications. The cause is partially that the receiver has reached task load saturation, and cannot continue to process information coming in. Another possibility is that maybe the receiver has too many filters in place, especially personal feelings about the sender, to really listen to the information. Remember! When we listen, our objective is to try as hard as we can to match the sender's perception of reality, not filter it to ours (see Fig. 5–1).

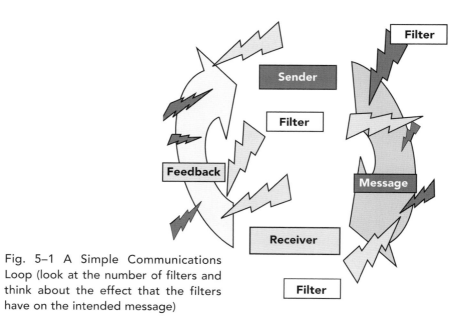

Fig. 5–1 A Simple Communications Loop (look at the number of filters and think about the effect that the filters have on the intended message)

FILTERS

Filters in communication affect how we receive the messages and information available to us. We begin to form our filters not long after we are born. Some items that affect our ability to communicate and listen effectively are race, religion, gender, education, language, experience, previous interactions (both with the sender of the message and others in our lifetime), technical abilities, where we grew up, family life, etc.

So, as you can see, filters are a reality and you must be aware of your own and attempt to navigate filters that others have in place. Sometimes filters are unintentional and at times filters are intentional. For example, a sender has no control over the filters that form by being a woman in the fire service. But she will have control over the filters that she has formed because many years ago she once had an officer who treated her with no respect. Without controlling the filters of previous discrimination, effective communication can be limited and feedback is not based solely on the situation at hand. There are thousands and thousands of filters that we communicate through every day. We need to be aware of them, use them effectively, and realize that not all of them will keep us safe and make us an effective team.

Conflict resolution

Conflict is a normal part of group interaction. All personnel in the team must expect that conflict will occur, even in highly organized and effective teams. The number one item to remember is, "What is right, not who is right." Respectful interaction and rational thinking, void of any inappropriate influences (such as race, gender, culture, religion, personal feelings, etc.), will lead to a successful resolution to any conflict.

A man was speaking of his neighbor and the problems they have being across the fence from one another. The man says, "I can't win! Every time we get in a disagreement he gets historical." The man's friend corrects him and says, "Don't you mean hysterical?" The man says, "No. I mean historical. He keeps bringing up stuff from 20 years ago." The key to conflict resolution is to focus on the issue at hand and not to bring in yesterday's dirty laundry.

During conflict management, emotions usually run high, even in experienced teams. Remember, passion and ability is what brought you all together, and that is what will make you successful in the future—but it will also create conflict. Conflict is necessary to build a strong team. Leadership and communication will work through the conflict and make the team stronger. In fact, teams that experience conflict early in their formation and work through it, will bond earlier and stronger than teams who conflict later in the formation process or not at all. A homogenous group, a group that cannot stand conflict, or groups that cannot work through their conflict are doomed to fail when the chips are down. Maintain control of the situation, and if possible schedule a time to discuss the issue after everyone has had a chance to cool off.

Do not allow other team members to be dragged into the anger. Concentrate on the issue and do not deal with personnel issues or personal issues in public. It is very foreseeable that problems formed outside the work environment may have entered into the equation and caused the problem. Each side should have an opportunity to calmly state its views and issues. This is a chance to clear up the misunderstandings that are usually the tiny seeds that (with constant watering) turn out to be a full-fledged cactus under someone's saddle blanket.

Each participant must maintain respect and strive to maintain respect of the other members. Separate the feelings from the facts. Always try to reach a solution that is workable for all participants. Unresolved conflict has torn apart many teams and driven off many more valuable team members.

Feedback

Proper communications depend on continuous feedback. It begins long before responding to an incident and continues until everyone is home. We must continuously provide feedback to decisions, actions, and incidents to improve safety and effectiveness. If we do not receive feedback, or if we get the routine head nod, go fishing for the proper feedback response. Have firefighters repeat the instructions back to you or ask them to explain a different method of doing it. Do not accept the *brown nosing* feedback, especially to important emergency communications. Simple, concise read backs will clear up a lot of misunderstandings and potentially dangerous situations.

The system in use

You are on a wildland fire and you notice a large column of smoke just over a small rise. The terrain and the wind could push that fire right up to where you and your partner have your truck. You call your division and ask what is happening. *(inquiry)*

Division states that they started a blacklining operation. You speak up and tell the division that you are between the main fire and their set fire. *(advocacy)*

Division says they are not sure where you are, but that you should be all right.

You state that you are in a difficult spot and the fire is going to come your way due to the terrain and the winds. *(advocacy)*

Division still says that it should probably be all right.

You state that you feel that if you stay there you will be trapped due to the access and the fire behavior; and that you will be pulling back to a safe area. *(advocacy, self-directive)*

After you state that you are leaving your assignment, it finally sinks in to the division that you are not comfortable. They tell you that they are stopping the blackline operation, and that you should be all right to stay where you are.

You agree and stay. Fifteen minutes later you still see a large column of smoke. *(monitoring)* You call division and ask if they can see it and what is going on. *(inquiry, without undue influence from the last inquiry/advocacy cycle)*

They state that they just had to finish up this little corner and they are almost done. You get into your truck and move your crew to a safer location where you contact your branch or operations director for assistance, *(advocacy, safety issue, violation of SOPs—lookouts, communications, escape routes, safety zones, begin conflict resolution steps)*.

So, What Is Effective Communication?

- Effective communication includes more than just one form of communication (verbal, non-verbal, symbolic, written) into a message and makes those styles match as closely as possible.

- Effective communication takes place when both the sender and receiver recognize that perceptions, influences, situations, and filters affect the message.

- Effective communications must have active listeners for comprehension.

- Effective communication uses the communication loop of sending, receiving, and feedback with all parties (see Fig. 5–2).

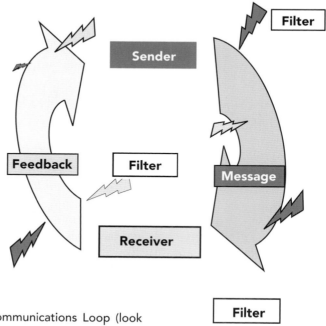

Fig. 5–2 Effective Communications Loop (look at the reduction in the amount and influence of filters on the intended message)

In one study, professional peers rated pilots "above average," when they used these concepts:

1. A first officer should be encouraged to question a captain and procedures.

2. Pilots should make their personal problems known to other crew members.

3. First officers may assume control of an aircraft in situations other than total incapacitation of the captain.

4. The pilot flying should verbalize plans and maneuvers and make sure that they are acknowledged and understood.[17]

This study could be replicated in the fire service, and the results might look something like this:

1. A firefighter should be encouraged to use inquiry/advocacy regarding an officer and that officer's decisions/actions.

2. Firefighters and officers alike should communicate any problems concerning physical health and/or mental health that could affect the safety of themselves or their crew.

3. Firefighters should be involved in the decisions and actions of their crew in situations other than officer incapacitation.

4. The officer should make the strategy and tactics known and understood to team members.

Because we operate under unusual amounts of stress in our duties every day, we need to make sure that we:

1. Create an open atmosphere where respectful communication is expected on and off the emergency incident. Good communications equal increased safety and more effective operations.

2. Understand that stress affects the way we communicate—the words we use, how we use them, and how they are understood.

3. Think before speaking. Do not get caught using a stock phrase in place of an actual, original thought.

4. It's not who's right; it's what's right.

5. It might not be wrong. It might just be different.

Use the concepts in these pages and your safety and efficiency will improve, but so will other areas, too. By communicating information constantly in a valid, workable system, you move part of your formal training program to an informal, on-the-job program. You increase mentoring opportunities and increase understanding between all members of the department—those getting the strategy and those giving the strategy. You'll actually transfer the tools of good fireground planning and initiation from those who know (the officers) to those who want to know (the firefighters), and bring those new personnel on board quicker and easier than you could have done in a more formal situation such as in the typical classroom settings we currently use.

REFERENCES

[1] Brunacini, A., *Fire Command*, National Fire Protection Association, Quincy, Mass., 1985. (Reprinted with permission from the National Fire Protection Association, Quincy, MA 02269, © 1985)

[2] Billings, C.E., and E. S.Cheaney (eds.), "Information transfer problems in the aviation system, (NASA TP-1875), Moffett Field, CA: NASA-Ames Research Center." From: *Sitting in the Hot Seat: Leaders and Teams for Critical Incident Management*. Rhona Flin, John Wiley, and Sons: 1996.

[3.] Sexton, J. B., and R. Helmreich, "Analyzing cockpit communication: The links between language, performance, error, and workload." University of Texas Human Factors Research Project, Austin, Texas. In Jensen, R., B. Cox, J. Callister, R. Lavis, (eds.) *Tenth International Symposium on Aviation Psychology*, Columbus: Ohio State University, 1999: 689.

[4.] Reprinted from: Helmreich, R. and C. Foushee, "Why Crew Resource Management? Empirical and theoretical bases of human factors training in aviation." In Wiener, E., B. Kanki, and R. Helmreich (eds.), *Cockpit Resource Management*, New York: Academic Press, 1993, with permission from Elsevier.

[5.] Dimitrius, J.E., and M. Mazzarella, *Reading People—How to understand people and predict their behavior—anytime, anyplace*, New York: Ballantine Books, 1998, 1999.

[6.] Dietrich, R., and D. Silberstein, *Group Interaction in High Risk Environments*, Sponsored by the Gottlieb Daimler- and Harl Benz-Foundation, Final Report of the Subproject: "Initiating Crew Resources under High Cognitive Workload," Berlin, 2001: 30–31.

[7.] Adams, S., *The Dilbert Principle*, New York: HarperCollins Publishers, Inc., 1996.

[8.] Dietrich, R., and D. Silberstein, *Group Interaction in High Risk Environments*, Sponsored by the Gottlieb Daimler- and Harl Benz-Foundation, Final Report of the Subproject: "Initiating Crew Resources under High Cognitive Workload," Berlin, 2001: 27.

[9.] Linde, C., "The quantitative study of communicative success: Politeness and accidents in aviation discourse." *Language in Society*, 17, 1988: 375–399.

[10.] Dietrich, R., and D. Silberstein, supra at page 9.

[11.] Ruffell Smith, H. P. "A simulator study of the interaction of pilot workload with errors, vigilance, and decisions." NASA Technical Memorandum 78482. Moffett Field, CA: NASA-Ames Research Center.1979. In J. Bryan Sexton and Robert L. Helmreich. Analyzing cockpit communication: The links between language, performance, error, and workload. University of Texas Team Research Project, Austin, Texas. In Jensen, R., Cox, B, Callister, J., Lavis, R, (eds.) *Tenth International Symposium on Aviation Psychology*, Columbus: Ohio State University, 1999: 689.

[12.] Chidester, T. R., R. L. Helmreich, S.E. Gregorich, and C. Geis, "Pilot personality and crew coordination: Implications for training and selection." *International Journal of Aviation Psychology*, 1, 1990: 23–42, from Sexton and Helmreich, supra.

[13.] Discussion Notes: Lynne Martin, PhD, University of Scotland, Aberdeen. *Ninth International Symposium of Aviation Psychology.* (Dr. Martin was kind enough to share her ideas in person, and not in a published paper).

[14.] Harrington, D. K., and C. D. Gaddy, "Overcoming the effects of acute stress through good team work practices." In Transactions of the ANS/ENS 1992 International Conference. La Grange Park, IL: American Nuclear Society, 1992: 86–87.

[15.] Tactic demonstrated by Scott Windisch, Ponderosa Volunteer Fire Department, TX, to Campbell County Fire Department, WY, Firefighters on scene of a structure fire.

[16.] "Analyzing cockpit communication: The links between language, performance, error, and workload." J. Bryan Sexton and Robert L. Helmreich. University of Texas Team Research Project, Austin, Texas; 2002.

[17.] Helmreich, R. L., H.C. Foushee, R. Benson, and R. Russini, R., "Cockpit management attitudes: Exploring the attitude–performance linkage." *Aviation, Space and Environmental Medicine*, 57, 1986:1198–2000.

IT'S CRM LEADERSHIP

SINCE THE WORD *leadership* appeared in the 1800s, literally tons of material has been written about it, and billions of dollars spent researching it. Although actual leadership has probably been around since two-celled organisms began, it is within the last 50 years that the perceptions of leadership have changed dramatically.

Within the last 20 years, leadership has changed even more. Sun Tzu, an ancient Chinese War Lord attained notoriety in the fire service when his book, *The Art of War*, was applied to fireground planning and operations. What he could attribute to experience, cunning, intelligence, intuitive leadership, good judgment, or just pure, hidden luck and chance, we must now study, justify, budget, track, certify, and prove. We sometimes forget that leadership is about people; it is management that is about things. Sun Tzu knew this thousands of years ago; today we relearn many of his lessons the hard way.

Today, because of the situations we put our leaders in, we must rely more and more on a team of leaders rather than a single leader. This is the leadership system that CRM helps to build. CRM and CRM leadership are not magic bullets sent from the future. Great leaders have been using these principles all along. The problem is, unless you were within the circle of influence of that great leader, there was a long and difficult road ahead to learn the concepts of leading people. CRM Leadership will help you be a great leader, and it will help you make your followers into great leaders, too.

The fire service does not need a bunch of New Age, warm fuzzies running around hugging everybody. That is not what CRM is about. The fire service needs individuals who can communicate, listen, and realize that humans are not infallible—they need help—we need help.

Within this chapter we will stray from the traditional (or non-traditional) discussions of leadership in the fire service. I am certain we can all agree that we would all rather be pounding our thumbs with hammers, than read through one more sentence on what traits make good leaders and their function in the fire service. We can sit down and draft a fairly exhaustive list of the traits that we find personally beneficial in a leader, and although it will be similar to a list drafted by another firefighter, it will not be exactly the same. The one problem in identifying and training leaders is that not everyone agrees on the definition of a good leader and the definitions change with the situation. Each department has its priorities and influences; it is not for us to create a punch-list of *must haves* to complete before assuming a leadership position. This being generally true, I think we can all agree on a few skills that a fire service leader should possess.

Good fire service leaders should be

- Technically Competent
- Honest
- Open

- Experienced
- Trained
- Knowledgeable
- Fair
- Mission- and Vision-minded
- Good Communicators
- Organized

You might write down other traits that you expect of your leaders in the margin; this might help you focus on this chapter. From here on we focus not so much on the actual person and the traits, as we will focus on the role in which they perform their duties and their responsibilities under CRM Leadership.

We do not intend to discount the countless hours and pages of leadership and management material out there. We believe that good leadership abilities come from a proper base of leadership understanding and management understanding. We tend to agree that there are quite enough of these existing courses available, and we do not want to duplicate or "put a spin" on existing material. Many, many excellent texts and courses exist that will give you the foundation you will need for continuing forward into CRM Leadership.

As with CRM as a whole, you must first gain technical proficiency in your area before you can hope to succeed as a CRM firefighter or leader. Once you have gained your base knowledge about the theories of leadership and management, the application of those theories, and the requirements within your department, then you can begin to apply the CRM concepts fully. Within this chapter, we discuss how to use CRM Leadership concepts and skills to make your operations safer and more effective, and to increase morale within your core groups.

Firefighters tend to follow three basic paths when they enter or are involved in the fire service. These paths may cross many times during a career. This over-simplification may seem pessimistic to some firefighters and leaders, but it is important that we all realize that firefighters follow their leaders for a variety of reasons. From this point-of-view, leaders can make the best decisions on how to lead their firefighters and build their teams.

Here are the three basic reasons firefighters follow their leaders:

1. Followers do not know any better and they feel compelled either through cultural, personal, peer, or organizational pressures to obey commands given by leaders and/or more senior personnel.

2. Followers know better, through training and experience, but still feel compelled to obey commands either because of cultural, personal, peer, or organizational pressures.

3. Followers truly have proof to trust their leader's abilities and skills.

These categories also loosely conform to the patterns of assimilation into an organization, occupation, and team. When personnel are first brought into an organization, they usually know little about the actual job and people. They are inundated by policies, procedures, acquaintances, and training. They bring with them the assumptions of the popular (and sometimes unpopular) media world in which we live. These include everything from Hollywood cars exploding on impact, instant death for gunfighters, everyday superhuman feats, the 30-minute television program resolution, and the clip-it news channels that follow real-life drama. These constantly add the spin that makes stories interesting to the non-firefighter audience.

A big part of the Baby Boomer Generation's background was shaped by this information blitz and the effects of a military climate. This reflects a climate where war, conflict, and the military were in the movies, in the news, and in their lives. This military climate had a lot of bearing on how people believed their role of followers should be carried out, and how the role of the leader should be accomplished. An order was an order and it was not to be questioned. The leader is the leader, and the follower is a mindless minion. Communications pathways were not bi-directional; a leader's decision was usually not open for discussion. In addressing CRM leadership, we need to make sure these people know that it is acceptable and expected that they contribute to the team by not only following orders, but also contributing their experience and training through communications, and improving the safety and effectiveness of an operation.

The new generation has grown up in an entirely different *media age*—these Generation Xers. A lot of management books and concepts have already been drafted that will teach us how to deal with these rebellious kids and all the challenges they bring with them. Once in a while, usually in the fall, an e-mail hits the Internet that is evidently aimed at trying to help college professors understand the next class of freshmen. It is interesting to read all of the things that the new class has never had, seen, or dealt with in their many, many years of life (18 or 19, I mean). These include things older generations believe everyone knows about, such as black and white television, vinyl LPs, non-electronic data bases (filing cabinets), and men on the moon. The perspective explains why new firefighters sometimes do and say the things they do—a different world equals different values.

One of the values that the newer generation has that shocks and appalls many others is talking back. Members of the newer generations do not have a problem (as a whole) in questioning a superior. (In the fire service we tend to beat that out of them the minute they arrive.) The new generation already has a head start on the concepts of CRM through the background information that surrounds them growing up. We need to harness that background, explain how to properly use it, when to use it, and support the use of their background abilities to make the emergency scene safer and more effective.

THE FIRST STEP—INTRODUCTION

Most rookies have little or no knowledge of actual operations, strategies, and tactics of the modern fire service. They follow their leaders because

1. that is what they think should be done; and

2. that is what their peers do or say to do; or

3. that is what the organization says to do. (Members are expected to follow all commands of their officer or other superiors unless said command is a violation of applicable laws or ethical standards.)

The old phrase, "You don't know what you don't know," applies here. It explains why these new firefighters feel compelled to follow the commands of someone who seems to know what the heck is going on—right or wrong. This mindless following is the first step in their assimilation into the organization. People who "talk back," don't usually make it very far in the organization, especially when they are brand new. Shut up and do as I say is usually the order of business. The louder the superior, the more he or she must know, right?

It is uncommon for individuals to remain at this unknowledgeable level; they usually enter as moldable clay, and through their time and efforts are at least marked up a bit, and enter the next stage of leadership/followership. As leaders, we can transfer the necessary technical knowledge that they need to perform their jobs, as well as instill in them the CRM Communications Principles of Inquiry and Advocacy. This gives them the skills they need to clarify what they don't understand and begin to form the foundations for their positions within the team.

Second Step—Integration

People with a general knowledge of the operations and policies of the organization understand why we are or why we are not performing certain operations. Much of the time they do not openly question the effectiveness (or sometimes even safety) of a command by a superior (although they may discuss their feelings within their team). For many reasons, this stage is the destination for many firefighters—although they will tell you with conviction that they are not here at all but at the third and final step.

The reasons they have not entirely made the leap to the next step is partially their fault, partially their leader's fault, and partially the organization's fault. When people are at this stage, they are given an order, good or bad, and they usually follow it. On occasion they may question the safety of an order. But for the most part, they are blindly following their superiors because they will tell you they "Trust them with their lives."

What is usually the case is that these individuals know when things are going well, or not so well, but just choose not to be a troublemaker by asking a lot of questions. They might say, "Besides, I just want to do my job and go home." Unfortunately, entrusting someone else with your life is not the best way to make sure you go home to your family. These individuals feel that the organization or the leader does not want feedback and that they have it, "all under control."

This impression can be based on either factual or implied information. If it is factual, something has happened to them personally or to someone else in the organization that proclaims speaking up to a superior is not allowed, tolerated, or even a good idea. If it is implied (which most of the time it is), that stubborn background comes back to haunt us. No one ever talked back to the *Duke* who was not looking up from the floor a second later, right?

People just do not feel that society affirms the success of people who do not follow orders. Guess what? Society usually does not. It is a tough pill to swallow when we admit that our background is incorrect for our present situation, and when we must change our innermost values and ideals. With support from yourself (that sounds silly doesn't it), your peers and leaders, and your organization, you can catapult your firefighters and yourself into the final stage of leadership/followership and be truly safe.

In the wildland fire community, there are Ten Standard Fire Orders that are supposed to be "unbendable and unbreakable." The father of the CRM movement in the fire service has developed the "Real Ten Standard Fire Orders."

Here are the official, published Fire Orders:

1. Fight fire aggressively but provide for safety first.

2. Initiate all action based on current and expected fire behavior.

3. Recognize current weather conditions and obtain forecasts.

4. Ensure instructions are given and understood.

5. Obtain current information on fire status.

6. Remain in communication with crew members, your supervisor, and adjoining forces.

7. Determine safety zones and escape routes.

8. Establish lookouts in potentially hazardous situations.

9. Retain control at all times.

10. Stay alert, keep calm, think clearly, act decisively.

The actual, on-the-ground application of those Fire Orders that is generally accepted, or at least allowed by the organizations, looks something like this:

1. Fight fire aggressively.

2. Maximize overtime.

3. Keep other costs down.

4. Promote self and crew image.

5. Promote agency image.

6. Shut up and butt out.

7. Don't say no.

8. Red card ratings are more important than experience.

9. Commit to an action and stay with it even if adverse changes occur.

10. Reporting safety infractions will adversely affect your career.

Although tongue-in-cheek, the *Real Fire Orders* show the repressive culture of the wildland fire community. In all honesty, I believe the structural fire service has similar problems. Whether factual or implied, these perceptions of expectations keep folks at this stage and contribute to about 100 firefighter deaths a year and hundreds of thousands of injuries.

THIRD STEP—TRUST

This last stage of truly trusting the leaders you follow is the summit of the mountain in organizations. Individuals at this stage truly *do*, "Trust their leaders with their lives." They do this not because someone told them to do that (remember, factual versus implied), or that they will not speak up, but because

1. the firefighters have the appropriate level of knowledge regarding their operations;

2. leaders share both information and lack of information with their team;

3. the team is allowed to share both information and their lack of information with their leaders; and

4. experience has proven this to be true.

Getting to the Third Step and Staying There

The first step is the easiest to complete. Simple training and teamwork, good, solid leadership, and motivated individuals at this stage make for a short time spent here. The second stage is more of a challenge to complete. Individuals can be here for a variety of reasons; some may be under the control of the leader and some not. But by applying these CRM Leadership Principles, most people will move beyond this step and spend a lot of time in the final step.

Two-way communications are a big part of this third step. Two-way communications (bi-directional) consist of the transfer of pertinent information and the transfer of the lack of information up and down the chain of command. The lack of information that a leader has is something that a follower really needs to be aware of. If the leader does not have a comfortable level of information, the team can gather it or adjust operations. Think about it in this example.

You are responding to a commercial address for a smoke report. Normal procedures are the same everywhere—investigate where the smoke is coming from, and make it stop coming from there. But, the leader says to himself, "Self, I'm not sure what Ex-Gener-Waste does in this building." If that is never transferred to the team, the team performs the standard operation and could possibly be at great risk. But if the leader says to the team, "Team, I don't know what Ex-Gener-Waste does in this building." The team will probably say to themselves, "Selves, maybe we should see what Ex-Gener-Waste does in this building before we rush in there." Maybe they take radioactive waste from power plants and make it into glow-in-the-dark lawn jockeys. *We don't know!* But simply stating that you do not have all the information you think you need for a sound decision, puts the team into the mode of watching out for everybody a little more carefully and gathering more information.

WHO IS THE LEADER?

Without a doubt you can define a leader as a person who holds a formal position within your organization. People such as company officers, captains, lieutenants, and chiefs are all leaders, formal leaders. Fortunately, most of the people in these positions are there for a reason. We may not always agree with their methods, decisions, and visions, but we are lucky to have dedicated individuals who are willing to serve in these positions. Unfortunately, there are some individuals in these positions who are not great leaders. They may be the best firefighters ever, but may lack in leadership and management skills. CRM Leadership will make those successful leaders more successful, and those marginal leaders much, much better.

Traditionally, most leadership lines of authority and channels of power travel up—not down. When the officer on-scene cannot effectively command the incident, or does not have the technical competency or knowledge, an officer on a higher level assumes control. Hardly ever, if at all, does the leadership authority travel down to a lower level in the organization, especially to a non-officer.

The fire service's notion that leaders must know everything about everything does not provide for today's complexities that firefighters face every day. Today's fire service leader knows how to build and maintain teams that get the job done—regardless of who is leading. Other forms of leadership in your organization are often overlooked and often under-supported or ignored. Individuals without formal rank or positions can many times be called upon to perform as leaders.

These people can come in different forms and serve many different roles. Below is a basic list of leadership roles and individuals, and how they can be employed to make the leader, the team, and the individual most successful.

Informal leader

Individuals who perform leadership activities without formal status on a continuing basis, could be classified as informal leaders. These individuals are often lead firefighters or experienced firefighters who take it upon themselves to orient others to their duties, tasks, and department values. Because of their status in the department, they are looked upon as individuals who can answer the questions posed, and are thus put into that role whether or not they enthusiastically accept.

It is important to clarify the informal leader status with respect to a person's willingness to lead. Those that have the status to lead but choose not to, or choose to lead others down the wrong path, must be monitored. The straying informal leader needs to know the effect such behavior might have on an individual and the organization, and if necessary be reined in.

An example of a positive informal leader would be a firefighter who has a lot of fire experience, lots of training, and supports the department's mission/vision. This individual may take it upon himself to offer his time to newer or inexperienced members to assist them in their duties. Every department seems to have an individual or individuals who operate like an "Underground Welcome Wagon" to the new people.

Opportunities abound for informal leadership. New people especially in volunteer systems enter into a complexity of social and professional systems. They need guidance and a friendly face to guide them through those intimidating first weeks. Those informal meetings often form a bond between people, and begin to transfer the group norms to the rookie. These informal one-on-one meetings and trainings are invaluable to the rookie and to the department. The department as a whole does not have to establish policies and procedures for these mentors. No one has to be in charge of assigning mentors. The mentors do not report to anyone on the progress of the rookie. Loosely speaking, I believe this system works because people are motivated, friendly, and outgoing; these are all traits of individuals in successful departments.

Departments sometimes try to mandate mentorship programs. Departments force individuals together like two north poles of a magnet. It just does not work like it should. People come together and form a bond because of many different reasons. Maybe it is similar interests or similar looks or similar gender or who knows what else.

To mandate that you and I are to be friends, and you are to come to me for any rookie questions, just is not as effective as the informal system, amazingly enough. If a mentorship program is important to you and your organization, do not mandate it. Do not force compliance or require participation. Work on the overall morale of the people who are already in your organization. Make it so they *want* to show people how proud they are to be affiliated with this group, and why the rookie should stay involved in it, too. That is the secret to mentorship—excited, professional, existing, experienced, informed (should I go on?) personnel.

Other opportunities for informal leadership exist outside the rookie example. During station duties, during responses, during the incident, during clean-up, and maybe even off-duty, we have situations crop up that open that door of leadership, and wait for us to push it open and walk through. Every individual on the team is the best at something. It may not be firefighting or medical or loading hose or knowing the town. But maybe that individual could make an engine run like brand new, or another could make patients more comfortable on a medical call than anyone we ever saw. When those two got together they could grab the attention of a class of eighth-graders like nobody's business. As a leader you should be identifying what people are best at. Find that and you have found their love and their opportunity for leadership. A simple skill used here and there in a leadership role begins the long process of building the leader of the future, a stronger team, and a more successful organization.

People who accept those informal leadership positions, for the most part, know it is temporary. They do not grab leadership like you grab the last donut in the box. They humbly accept it for what it is, a chance to show that they are best at something. This is an opportunity to shine and to use this small point in time to teach others something that they are passionate about. They accept this position, perform to the best of their abilities, and then return to their own position, hopefully feeling good about themselves and what they contributed. Now they have another piece of experience they can keep and build upon.

As the formal leader of the group, you have important steps to take before, during, and after an informal leader takes control of a situation.

1. **Know the individuals on your team.** You cannot create opportunities for informal leadership without knowing who your people are and their capabilities.

2. **Create mini-opportunities for informal leadership whenever possible.** Do not expect your people to jump into a leadership role on a six-alarm commercial fire without giving them some foundation first. Mini-opportunities are all around us at the station. ("Hey, Barb. Would you mind making sure that the Public Education Kits are up-to-date on the trucks? Why don't you take Jeff with you and show him what's in them?"). Or during training— ("Hey, Barb. Would you mind going over those new procedures on the truck with everyone?"). On the incident—("Hey, Barb. Take Ron, and make sure he knows how to identify the zone on those new fire alarm systems you learned about.") After the call—("Hey, Barb. Can you take the crew and make sure that the truck gets back in service while I finish this report?")

    This not only gives the individuals an opportunity to show their stuff, but it also proves that you are the kind of leader that does not hold onto command like a bull-rider hangs onto a bull. (You can throw 'em...sometimes, but it ain't easy.) Give up your role as leader once in a while, not just because, but because it improves your team and you. Maybe you could learn something, too; and at some point you will relinquish your role, either willingly or unwillingly, so you might as well plan ahead.

3. **Support those informal leaders.** Do not give them the assignment of making sure the guillotine is sharp enough; a one-time do or die opportunity is likely to ruin a good team member. These individuals are not going to be perfect. Hopefully, you have given them enough opportunities in non-stressful situations that they will perform at least adequately, when necessary. We cannot expect them to be great leaders or do things like we usually do.

    These people are in a new role and will perform differently. Do not step over a dollar to pick up a dime. Not supporting them now by berating them and their abilities, etc. is like picking up the dime that you see right now instead of working a little bit more to get the dollar. Support those efforts through suggestions during the action and critiques after the action. By incrementally supporting them each and every time, you will build an individual who will in time be able to perform as an informal leader seamlessly.

4. **After the fact, make sure everyone knows how you feel about informal leaders.** Be honest with the group. Do not serve them hamburger and call it prime rib. People know whether it was a good performance or a bad performance. The issue is not that the informal leader succeeded or failed, it is that he or she took the opportunity with your support. Work through the tough times and they will occur less and less often. Sugarcoat everything and pretty soon you are riding your fire truck naked down the street proclaiming to everyone how good you and your team look in their new clothes. It won't fool anyone for very long, if you are dishonest and your performance as an organization will not improve.

Situational leadership

Another type of leadership without formal position is the situational leader. These individuals have specific expertise that is useful under the right circumstances. These people may be junior officers, firefighters, inexperienced members, or even members who are not well respected. Their position as leader is established because of their knowledge or experience in a particular situation.

For example, a volunteer firefighter with lots of experience fighting fires and many hours of training had just begun a new job. Within the first two months he was called upon to be a situational leader at a fire on-site where he was newly employed. The fire was in a large coal storage shed.

Many factors existed that could prevent this firefighter from becoming a situational leader.

- He had little experience in the mining industry—he was newly employed.

- He was on a standard, new-hire, probationary status with the company.

- He was unfamiliar with many of the people he would have to lead; and

- Many layers of individuals with much more experience were present in the actual building where the fire was located, and knew the processes involved.

Fortunately there were many other factors that enabled this volunteer firefighter to become a situational leader.

- He had extensive background and training in firefighting, Incident Command System (ICS), and fire command.

- He had become acquainted with some key individuals who were familiar with his abilities and they supported him.

- He had fought a fire just like this one in another company while on a fire call, and the lessons were directly transferable.

This firefighter took over the command and control of this fire with the support of the company representatives and the site emergency response team. This situational leader worked with the company representatives to prevent the same mistakes that were made on the previous fire—many of which were more or less SOPs for the industry, and would have caused extensive damage to the structure due to the fire conditions. The previous experience in this type of fire had resulted in about $750,000 damage. Because of the lessons learned, the loss on the second fire was only $65,000! As you can see, sometimes the right leader for the job is the one who might be easily passed over when employing traditional channels of power and lines of authority.

With the fire service's repeated incursions into unfamiliar territory, the situational leader will become more and more important to the safety of the firefighters and the citizens, and to the effective resolution of the incident. Incidents in which the fire service lacks training, experience, and/or equipment will require use of situational leaders who have prior knowledge and experience, possibly from areas outside the fire service to deal with the incidents, creatively, effectively, and safely.

Even areas inside our scope of control may require the acceptance of the situational leadership role. Incidents such as technical rescue, hazardous materials, violent situations, terrorism, etc., are all examples of areas in which a mass of knowledge is available and required. But these are areas for which the fire service has had little time or money to spend to prepare its firefighters to deal with these newer areas effectively and safely.

WHAT SHOULD A LEADER BE DOING?

"It depends." That answer is a departure from other leadership courses in the fire service. Those courses spell out a fairly formal and orderly job description for those fire service leaders—as if managing piles of paper is almost like leading people. Realistically what the firefighters should do in the fire house, on the scene, after the scene, etc., can (and probably should) be different. Leaders remember, you can manage stuff, but people must be led.

In the fire service we have drawn many of our concepts of leadership from the military ideals of command and control. But what we often forget is that we are not the military. When the military commanders plan an operation they look at many different factors. They look at the battle within the context of the overall theater of operations and their main goal. (Will taking this hill have a positive effect on capturing this city?) They look at the resources available. (I have five platoons. Will that be enough to take this hill? Do I need other forms of support such as artillery or air?) Then the toughest factor is how many men will we lose to gain our objective and is it worth it? (Under current intelligence estimates, we should be able to capture this hill with the resources and time we have available with a loss of 15 men.)

Can you imagine planning a structural fire attack in this context? Let's see, we need to stop this one-room structure fire. I have three engines and a ladder company with a water supply. I think we can stop this fire, and we will probably only lose three firefighters. Unfortunately our risk-versus-gain calculations are often skewed toward taking more risk because of who we are, and the fact that we are already there and ready to act, than the actual gain that can be expected from our actions. Why do we sometimes plan and deploy our personnel in situations where the risk of death is too great? We need to change our expectations of ourselves and our teams on these incidents. We should not be going to these jobs with the goal of putting the fire out. We should be going to these jobs with the expectation, "Everyone Goes Home Safe."

Too Much, Too Late

If you are (or want to be) a fire service leader, and you think your leadership is only for use on the fireground, you are too late. You may have authority to command firefighters at a scene, but that is a far cry from having your firefighters' trust to lead them on-scene. A fatal example of this occurred on The South Canyon Fire in Colorado 1994. Firefighters from many different crews, with many different firefighting strategies and experiences, were brought together to battle what appeared to be a routine wildland fire. The leadership was arguably fairly strong within each crew (intra-crew), but very weak between the crews (inter-crew). This was a source of discussion between the personnel and probably was addressed, in a fashion, when crew leaders met.

Unfortunately the lack of leadership as a whole for the fire was not addressed formally. On the days before the blow-up, this problem was no more than a "rub" to the firefighters involved. But in the late afternoon of July 6, that rub became a major problem—specifically for two individuals of a helitack crew. When the fire made a run from the bottom of the canyon up the slope to the place where the firefighters were working, a decision was made to bail over the ridge into the drainage to escape the fire.

Although this small team of firefighters (the helitack) were ordered, requested, told—(whatever you would call this emergency communication) to follow the rest of the firefighters to safety, they chose a different route. That route ended in their deaths, while the others who dropped into the drainage opposite the blowup lived. The problem did not start and stop with a bad decision about where to survive the fire, or about whether to obey or disobey a direct order. The problem started a long time before, when the previous rubs and lack of leadership were not formally addressed, and when attempts failed to integrate the parts into the whole.

It is not the day-to-day boredom that defines the strength of a team, it is the seconds before, during, and after the event that causes adults to panic and become individuals—survivalists. Leadership, and not just formal leadership, begins when you first meet someone. It is not about formal positions, authority, and power for the most part, although firefighters will follow orders from their superiors. It is about bringing a group of firefighters together as a team. A team is a group, but a group is not necessarily a team. Therefore as a leader you should be forming your team—constantly. Also as a member of a team, you must strive to maintain the team. Simply bringing people together to form a group is not sufficient for those critical times, those nonstandard times. That is arguably what happened on the hill that day.

The helitack crewmembers had a lot of experience together and probably trusted each other. But they had little time to get to know the other firefighters on the fire. The leadership interface between the crews was rocky, and the overall fire leadership was lacking. There was no effort made to establish trust in the leadership on the day-to-day operations. Therefore when it came down to a live-or-die order from the leadership, the helitack crewmembers decided to go with what they knew and who they trusted—each other. Unfortunately their lack of trust in their leadership and their fellow firefighters, and a bad decision spelled their doom.

As a leader, you must accept your job to build the trust in yourself and within your team. This day-to-day trust will translate into a stronger team, and when the chips are down, and a snap decision must be made, your team will have an easier time reacting to an urgent command. They will do this because of the trust that has been established. Many times without the trust, they will spilt off from the group and go with their own decision or the decision of a more trusted and familiar team member.

WHY WORRY ABOUT BUILDING A TEAM?

Back to hard statistics. In the commercial aviation community: The NTSB studied 37 air carrier mishaps and found that 73% occurred on the first day the captain and the first officer were together, and 44% occurred on their first flight together.[2] These aviators are trained to highly professional standards, and are regulated by unbendable FAA rules and follow strict company SOPs. These are not folks with a ticket to fly who are randomly thrown together in the front of an airplane. These are the epitome of professionalism in a very professional occupation. Yet, they still crash—especially on their first day together because they have not come together as a team.

A majority of the time it will not really matter how strong or weak the team is because the situation will be standard with no problems or very simple problems. Performance will not be optimum, but most of the time it will be marginal enough to prevent notice. The issue is within a very minuscule amount of time when Murphy (as in Murphy's Law) throws you a knuckleball, and you absolutely must work as a team, and not just two people in the same space. That small amount of time marks when many of our firefighters are hurt or killed every year. They must work together within a strong team environment to ensure their safety and the safety of their team members.

TEAM PERFORMANCE ISSUES

Do you think teams work better after they have had time to form and work as a team for a while? A study of Naval Inflight Refueling Crews suggests, "Taken as a whole, crew familiarity research suggests that familiarity impacts coordination and performance positively on early sorties and negatively later on."[3] Implications on the fireground could be that your team is weak initially, gets stronger, and then performance deteriorates over time due to various factors.

Initially everyone is trying hard to work together, work through issues professionally, and learn from each other. It could be said that they are more open to the other team members in this first time period. Of course, this theory applies more in an organization that has strong leadership, adherence to SOPs, and loyalty. This is due in large part to the "roommate effect."

If you have ever had a roommate, you can vouch for this effect. The first week you are together is sometimes stressful and awkward; but nevertheless, you both strive to get along, setup and obey rules, and get to know one another. After a couple of months it is like "cabin fever" sets in and you cannot stand your roommate. Everything that roommate does drives you crazy. You know too much about that person and you may not like or even respect your roommate anymore. You must still live together, but it is often grudgingly. You do not hesitate to bring up the other's bad points when arguments arise over trivial items (like who left the dirty dishes in the sink). Fortunately, at home, an incident that requires teamwork is not too common. Unfortunately, in the fire service, it could be a daily exposure.

Teams must be built for the majority on a professional level and on a personal level. That is a must for trust, loyalty, and respect. What can cause problems in a team occurs when the majority of your teamwork is built on a personal or social level, (which is often less stable over time), and to a lesser extent on the professional level. In addition, teams built on a social level tend to exclude individuals who are perfectly capable professionally, but lack the necessary social stuff to "be one of the guys."

Take an example of a small department that was made up of about three-quarters motorcycle enthusiasts. Of course when the motor-cycle gangs began to congeal in the department there was no problem—it actually helped. People were bonding and becoming closer, which is a big part of team formation. Some of the gang were in management positions and they began to assign crews and duties with much forethought and planning. After all, if they get along so well together, why not just put them together permanently?

All was still well in this department even though there were the gang crews and the non-gang crews. But then as time progressed, owning a motorcycle became more of a prerequisite to being a member of the department than just a social coincidence. Soon the line was drawn in the sand; was it going to be gang or *free!* This department was never able to separate its social and professional life. Maybe it should or maybe it should not—that is not the question (or something to that effect). The point is to be careful when social activities and interests begin to play major roles in a professional situation. The emergency scene is no place for social or professional monomania.

When teams have been together for a while, there is the possibility that their performance will degrade below acceptable levels. A leader must watch for the warning signs and take actions to fix the team. Some warning signs of a dysfunctioning team could be

- The team breaking into several smaller groups as a norm rather than the exception

- Frequent disagreements and arguments

- Personal matters brought into professional discussions

- Open disrespect for team members, their values, and/or possessions

- Observed lack of teamwork and support

- Organizational inaction towards necessary SOPs, meeting standards, etc.

- Tardiness/absences

- Lack of motivation

- Destructive, behind-the-back gossip

Part of the problem with teamwork in the fire service is that it is often not a requirement established by the department. Firefighters are measured almost daily on a variety of other requirements. They are timed to see how fast they can put on bunker gear, how fast they can don SCBAs, or their thoroughness in searching rooms. They have their strength and flexibility measured. They are critiqued on their ability to pull hose, bed hose, roll hose, check hose—hose! hose! hose!

How often are firefighters tested for their teamwork abilities? How do we train teamwork? How do we provide remedial training for someone who does not exhibit the proper degree of teamwork? What is the proper degree of teamwork? These are all questions that CRM attempts to answer. It has not answered all of them yet, but it makes a pretty good case for how leaders get their people there and keep them there.

BUILDING THE TEAM

"...a cosmology episode [is] an interlude in which the orderliness of the universe is called into question because both understanding and procedures for sense making collapse together. People stop thinking and panic."[4]

During non-stressful times, firefighters will tell you that they, at least partially, would disregard their own safety for the safety of their partners or for the rescue of a civilian. But research has shown that when really bad things happen and it's a life or death situation, humans tend to react in predictable ways. They resort to prior, over-learned behavior, they lose the mental capacity for creativeness and cognitive processing, and they tend to separate from the group and attempt to save themselves on their own.

This is not something we like to think about, but we must. As leaders, we place our people in dangerous situations constantly. We attempt to plan for the safety of our people, but there are sometimes unknowns that can create a situation where the death of one of our own appears imminent. It is at these times that the strength of the team is paramount to the survival of the individual. Unfortunately the individual instinctively wants to "go it alone." So what can a leader do to maintain the team and improve their chances for survival? Start *now!*

The easy majority of the time, firefighters can do their jobs with little or no real teamwork. There is no need for team-based, creative problem-solving under the stress of life or death situations. But as leaders we should be preparing for those times when it will happen. We must not allow our trainings to amble on, concentrating just on technical competence with equipment and strict obedience of written policy. We must stress our people to the point where they must use all team members within a team context. No fragmentation into smaller groups or into individuals on their own must occur. We must plan, train, and test these people constantly to instill in them the necessity of remaining together and thinking creatively during times of stress.

I just recently did a search drill for a green squad of firefighters. During this time, a firefighter ran out of air while deep inside the structure. The search leader made it to a door only to find it was locked. They took off to find another exit. I stopped their travels and asked them what was behind the door. They thought there was a good possibility that there was a safer environment behind it (true, there was, good job). Unfortunately they only had one axe for their tools and it was a metal door (I'll bet that doesn't happen again). Therefore opening the door was improbable. But what they had not thought of (the creative thinking process) was that the wall adjacent to the door was sheetrock and wood studs—perfectly "axe-able."

What we need to train into firefighters is that everyone needs to be thinking creatively in life-and-death situations because the standard operations and patterns are probably out the window. We must be able to utilize all the information processing power on the team instead of just looking out for ourselves.

As we stated earlier, the current safety culture is missing a portion of the big picture. We need a movement that changes our number one priority in the fire service. What once was *lives, incident stabilization,* and *property conservation* does not address our most important resource on the fire ground—*us!* When asked what the number one priority should be, firefighters say, "safety," or "saving lives," or "putting out the fire," or "serving our customers."

We would often introduce to the firefighters when we first began CRM in the fire service that our number one priority is, "You go home safe." But after more research, training, and experience with CRM we adjusted it to say that, our number one priority in the fire service is to make sure, "Everyone goes home safe." By everyone, we mean all firefighters. This reflects the team atmosphere and camaraderie that firefighters feel for one another; and it builds upon Maslow's Hierarchy of Needs by placing a feeling of personal safety utmost in our minds and our jobs.

Imagine what a change it would make both in firefighter injuries and firefighter fatalities, if we changed our number one priorities from saving lives and fighting fire to, "Everyone goes home safe." If honestly implemented, that change could be the catalyst for preventing more than half of our yearly fatalities and injuries.

Concentrate on the strength and safety of your team as a first priority, not the operation. Current safety programs are ideal for teaching firefighters how not to get hurt when using an axe to chop a hole in a roof, or how to roll hose to prevent a back injury. But they are very ineffective when dealing with abstract problems, problems of perception, or problems that escape traditional explanation or definition—problems that all fall under the category of working with humans on dynamic, real world scenes.

Our current safety programs concentrate on safety within the structure of an incident. CRM takes safety out of the operational mode and allows firefighters to truly be safe. It gives them the tools and skills they need to stay safe as the number one priority—not the number one priority *while* doing a task. It may seem like a fine (and possibly indistinguishable) line, but it makes a big difference to the operation of the team. Is it okay to stop a task that is unsafe under current safety culture and training? *Yes!* Is it okay to stop an entire operation that is unsafe? Arguably, *No.* When planning an operation, you should make sure safety allows the operation to take place, and not how to safely perform the operation.

Begin the formation of your team with a single-minded vision of your duties and roles—"Everyone Goes Home Safe." We do not believe that the American fire service will ever see a year where we have zero accidents, zero injuries, and zero line-of-duty deaths. There are too many factors working against us to achieve that goal. It is nonetheless a goal that we should strive to achieve.

We know that the emergency services are an inherently dangerous and unpredictable occupation, and we know that responders will be killed and injured doing their best for their citizens. What we do not believe is that the current trend of cardiac deaths, driving-related deaths, deaths during operations on roadways, "stand and fight" mentalities, and the "lives for trees" program are acceptable. These are not noble or heroic deaths. They are sad and tragic deaths that literally change hundreds and maybe thousands of lives—some so distant to the actual person that they have never even heard their name spoken, much less the kind memorializations. They stress and strain their coworkers and friends, place unbearable pains and difficulties on their families, and cause undue stressors on the systems that supported them. The smart firefighter may leave a battle today so that he or she may fight the war again tomorrow.

Your function as a leader is to form your team for strength during the emergency incident. We must be careful to differentiate our fire-fighting teams from those groups/teams that have become so popular in the business world. While we as firefighters strive for perfection, our focus should not be solely on the quality of our product or the lack of errors, but on our people and their safety and effectiveness. The emergency scene is not the place for management fads that preach zero-defect performance through team processes.

"By applying team-oriented, consensus based processes constantly, our leaders lose the ability to read the situation and act accordingly."[5] These processes do not allow our leaders to become skilled at applying information during dynamic events—often making life and death decisions with only 40% of the information. Nor do these methods reinforce the leader's responsibility for absolute authority and responsibility over actions and outcomes. As a formal leader, you must take special precautions against giving away too much responsibility and/or authority to your team or group processes. As the formal leader, you are the one who must have the final say, even in the presence of informal or situational leadership.

But it is also you who must accept feedback from your team and respect and value their contributions to you as if they were a gift. It is also you who must bear the lifetime of remorse if you ignore your team to their detriment. Use your authority sparingly, but use it when necessary.

Having made that point, I will say that building a team either through long-term assignments or short-term duties is the single most important job that you have. It encompasses copious amounts of knowledge, psychology, sociology, special skills, communication strategies, values, etc. These following pages will hopefully give you the basis for applying CRM team-building strategies to improve your operations and safety.

Shared mental models

To be successful, teams must be organized and supported correctly. CRM details many items that assist leaders and team members in this regard. Team members must possess a shared mental model of their roles and duties. Training, planning, briefings, and communications help firefighters realize how they support the operation as a whole and what their goals are in the operation itself. Without a shared mental model of the overall goal and the individual strategies, the team cannot hope to succeed in difficult emergency situations.

Clear direction

Along with the shared mental model of the situation, teams must be given a clear course of action. Ambiguous orders, lack of information, and other problems deny firefighters the right to a safe assignment and it denies the public the right to a good service. Make sure the team is given clear direction toward the shared mental model and then give them the support they should have.

Consistency

One of the most common and desirable traits in an organization and its leadership is consistency. So it goes with the members of a team. Knowing what to expect is vital to the success and maintenance of a team. The dependability of team members to perform their duties and support the shared mental model is paramount in the cohesion of the team. Inconsistency in leadership, organizational values, and other areas of uncertainty develops into a fragmented, unsuccessful, unsafe team.

Rewards

Firefighters do not expect medals, ribbons, and donuts for every fire they extinguish, but a simple rewarding word is very valuable. Without positive reinforcement of good behavior, that behavior does not seem to be valued. Likewise negative behavior should not be rewarded or tolerated. Again a simple word may be all that is necessary to correct the action and reinforce the expectations of the team or team member. Use the reward system correctly. Rewarding marginal behavior is almost as bad as rewarding bad behavior. Be honest and employ good interpersonal skills to reinforce the correct behaviors, and tell the firefighters how much they mean to the team and the community.

THE ORGANIZATION'S ROLE IN TEAM-BUILDING

Your department's administrative structure has a major role in the success of your teams—especially newly formed or extemporaneous teams. Many agencies, both public and private, attempt to assist the fire service through rules, regulations, and standards. What few of those agencies actually realize is that many of our emergency work teams are extemporaneous in nature. This means that the teams are loosely—if at all—organized. In volunteer and combination departments, your peer group may be 10 to 500, while your operational team is the standard two to six. Your team is rarely ever a repeat of the same two to six individuals.

This situation is a big difference from full-time, paid services where there is a set schedule for a team to work together, and while there are occasional adjustments, the normal team is together on most incidents. While attempting to keep everyone safe through these standards and regulations, the administrators often are detrimental to the actual operation and safety of the department—especially smaller volunteer services. These services become overwhelmed by the volume and compliance costs of standards and regulations. They have neither the manpower, the resources, nor the budget to maintain a favorable level within those standards and rules for any length of time. There are those departments that do not raise enough money to put fuel in their 1950s era, donated trucks to fight fire on their 500-square mile protection area, much less enough money for meeting all the current regulations and standards.

Therefore within your organization there must be a conscious acknowledgment made of the available, pertinent standards, regulations, rules, and accepted operating practices. A conscious decision must be made as to the applicability of the material to your department and situation, the ability to meet the requirements of that material, and the effect of that decision on the community, the department, its leaders, and its firefighters. Then the department must support its decisions to either follow or disregard the material through SOPs and local best-work practices and funding. The implications for not following the requirements and standards must be acknowledged. A department that applies lip service to the standards and to their members are setting themselves up for a problem when it comes to building and maintaining their teams and their safety. Here is an example:

A small, rural, volunteer fire department has just been handed two new books from the state fire agency. One is the new edition of the Basics of Firefighting from IFSTA; the other is a new standard for health testing and physical fitness from NFPA. No more money has been added to the budget nationally, regionally, or locally. The new training manual is easy to read and the skills are well described with photographs that demonstrate the skill—let us say that the skill is a new way to perform a room search. The method is quite a bit different than the way they had searched a room before, but most of them accept the new methods, and they begin training for it on their regular, once-a-month, drill night.

In a couple of months, most of them are pretty good at it and the department's organization supports the new methods. On the other hand, the new health and fitness standard is very difficult to read, often written in jargon that is unknown to a non-academic in those fields. The standard is long and very in-depth.

When the volunteer chief of the department begins the budget process, he thinks back to this standard, and since the national and regional authorities support it (mostly through lip-service), he assumes that he must also support it. He takes his pencil and writes in the number of members in his department, "25." Then he writes down the price for complying, "$1,900 each." The department's normal budget to run its donated truck fleet is $2,900 per year. It is clear, the health and fitness standard will not be approved.

This department was sent mixed signals from the governing bodies and standards agencies. On one hand, the department is surrounded by the new technologies and hazards. They are given all the information to train for the "new response" and it is very cheap most of the time just as in this example. They did not have to purchase anything; all they had to give was their time and minds, very simple for any department and any firefighter. The health and fitness standards are very important. About half of our line-of-duty deaths each year are from cardiac causes and at least half of those are known problems. With proper physicals, the majority of those cardiac problems would be known, treated correctly, and the firefighter's activities and duties adjusted. But the price of implementation is too high.

There is not much the department can do except imply that these standards should be met during the individuals private health exams. Unfortunately the simple solution changed the way the firefighters in this department performed searches, but not their physical readiness to engage in the skill. In fact, they could be more at risk now.

As an organization, we must make some tough choices for the safety of the team. A level of service must be equated to the actual level of compliance, not the implied level of compliance. The new search protocol might not be implemented, if we cannot assure that our firefighters meet the applicable health and fitness levels. What if a new pumper is purchased that doubles the available capacity of the previous unit, and provides a dependable unit with which to perform interior operations, but the department only has three SCBAs? What does this mean to the customers? The firefighters? Should they violate the "two-in, two-out" regulation in the name of service and effectiveness?

Departments must perform only within their levels and abilities, regardless of pressures inside or outside their organization. If all they are able to do is to show up with two people and spray water through the window, that is the level of service that is available. If they are a highly trained, well-equipped department that responds with 35 people, their level of service may be to aggressively attack the fire through the interior. But what if that department had a water system that only supplied a portion of the water needed? Could they still provide the same level of service? Many factors influence a department's level of service. The most important service we can provide is not to our public. It is to make sure, "Everyone Goes Home Safe," and by everyone we mean firefighters.

To support teamwork and ensure its success, an organization should

1. Identify the level of service that meets its unique local situation and actual—*not* implied, compliance levels.

2. Provide verbal and written information for firefighters to follow in support of these decisions.

3. Provide continuing assistance and support for methods of attaining and maintaining the levels of service.

4. Provide reactions and responses to efforts and actions that do, or do not, support the agreed upon levels of service.

All of these items can be accomplished through budgetary measures, written communications, discussions between the organization's decision-makers and the firefighters, through SOPs, and a strong , practical training program.

TEAMS:
THE LEADER'S RESPONSIBILITIES

The initial meeting

Hopefully your initial meeting will not be held during an emergency incident, but there are many departments and situations where this will occur. Federal wildland fires are a prime example, as well as any mutual aid responses. To ensure the safety of the team and effective operations, the organization hopefully has some pretty strong and standardized training. This training will need to take the place of the experience and cohesion that you lack as a team.

If you are lucky enough to meet your future team members prior to an incident, the size-up begins and the bonds begin to form. There are the formal items that can begin your team-building such as planning for it, training, exercises, written items, etc. There are officers who assume control of the group and the responsibility for its cohesion and success. They loosely control the situation in which the team meets, and what is discussed. Informally, the human animal is sizing up its next victim. We are looking for any weaknesses, strengths, mannerisms, appearance, etc. Not that we intentionally do this, it's just a natural ability and habit to size-up our new acquaintances. Dogs bare their teeth and fight, humans ask, "So, where you from?"

As a formal leader, you can take the lead during this initial meeting and make some giant steps toward building your team. The first meeting can either set the stage for a promising future or can damage the structure enough that it will take a lot of time to repair. Again team formation and human interaction is a very subjective set of skills. You will learn and re-learn from your experiences your whole life. Some simple things will help make the planned initial meeting a success

1. **Prepare for the meeting if possible.** Make sure you promote the image you want to reinforce in your firefighters. For some, this is casual dress and surroundings, while others are more restrictive in their dress codes and meeting areas. Whatever your culture is, ensure the meeting area is suitable. Ensure necessary materials and required items are available. Communicate any prerequisites to the team members so that they too may arrive prepared.

2. **Introduce yourself and your position.** This can be formal or informal as the situation requires. Some departments are very restrictive about how they address superiors and subordinates; others are more informal. How you address yourself will set the stage for the meeting. Requesting that you be called, "Captain Jones" will make the meeting more formal, while asking the new member to call you "Booger" will allow that a more informal tone is acceptable.

3. **Make sure everyone is introduced and has a chance to make a statement to the group.** This is the standard class opening stuff. Name, where are you from, why are you here, didn't I see you last night on the Ten Most Wanted List? It is a guided introduction that temporarily bonds the group, and makes them more open to other items and activities.

4. **Take care of administrative requirements.** During an emergency this may mean pulling out the certification card from your wallets to help identify—officially—the levels of training inside your team.

5. **The leader should discuss the expectations for the team and the team members.** This is very important. If the department is mostly extemporaneous in team formation, the expectations should be sufficiently broad to cover a variety of

situations and leadership styles. If the team is going to be together for a while, the discussion will set the standards and measures for the duration of the team in all they may be exposed to in their duties.

The initial meeting is important. It can be as long or short as the situation allows. But there should be some attempt made to connect with the new team member either through professional or personal channels, and opportunities and activities should exist for team members to begin to bond. This will promote the assumption of the team member or group of members into the next stage of team formation.

The new team

After the initial meeting, the work really begins. Sociologist Jon Driessen, Ph.D., studied seasonal workers in the Forest Service and determined that is takes about 6–8 weeks for them to "click" into cohesive teams.[6] In that time, leaders need to have a definite plan that they follow to hasten the formation of a strong team. In the Federal Wildland Community, crews are brought together in both classroom and physical training, sometimes weeks before they will ever be expected to respond to a fire call. This gives the leaders and the team time to form or gel and become a cohesive unit.

Structured and unstructured activities will help to develop your group into a team. Structured activities will give everyone a chance to begin to learn the acceptable norms and values of the organization, the leader, and the team. Structured activities should include

- **Classroom training.** Items to discuss are department SOPs, standard seat assignments, expectations of team members, transfer of pertinent knowledge, etc. This can also help the student understand the local fire problems, construction types/hazards, area streets, fuel types, frequent flyers, strategies and tactics, etc.

- **Scenario tabletops.** By burning down buildings or trees in the classroom, as a leader, you will transfer priceless knowledge to your team members. Every command officer has a personal style and presence; it is difficult to transfer this information in a traditional classroom lecture. But in a scenario, students may begin to learn what is expected of them, how the leader will command, etc. This helps to develop the concept of shared mental models. After a few scenarios, students will begin to think ahead, and expect the next decision/action of the leader. They will expect information from the leader at certain points, and they will offer information at strategic points without being queried.

  Overall, scenarios are one of the best ways to speed team formation to an end result of a fully functional and effective team. Of course, the more realistic the scenario, the more profitable it is from a training standpoint. I am certain we have been involved in those scenarios that are so outrageous that the only information transfer taking place is the coffee pot chatter about what lunatic organized that session. Make it real and allow the information to flow back and forth.

- **Physical Training.** Fitness levels are a high priority in the emergency services. They increase personal health, reduce health costs, increase operational effectiveness, and decrease injuries and fatalities. Time should be established on every shift, for team members to work out. This establishes physical fitness as a norm for the group. Team members who are below the acceptable levels should be coached to improve their levels—although most will not want to be left behind, and will vigorously pursue the level independently.

  Physical training can be very effective in forming your team. The physical challenge, exhaustion, mental strain and toughness, and completion of a difficult physical fitness challenge can immediately bring those people who finish it together as a team. A bonding takes place because of the stressful and uncertain circumstances of the effort. Many disciplines employ this technique of team formation: the armed services, various emergency service academies, elite wildland firefighters, etc. Leaders will need to participate with their teams and should exemplify the fit firefighter. Nothing is worse than a fat doctor telling you to lose weight, right? Do as I say, not as I do will not work with adults.

- **Skills training.** Training on the actual skills necessary for the job will provide an opportunity for members to help each other and learn what is expected of them—specifically from the leader. SCBA drills, hose evolutions, tool use, etc. are all important parts of the team's job, and the technical proficiency can help your team develop trust in each other.

- **Reality.** Real incidents are the ultimate test of the team. Debriefing (or critiquing) of an incident is very important to the formation of the team, and the continued effectiveness of operations. The debriefing chapter later in this book will describe the essential CRM elements of the debriefing, how to conduct one, etc.

- **Unstructured time.** Of course these people are not kinder-gartners; they do not need 24/7 supervision and control. Used wisely, unstructured time will help form bonds that may or may not carry over onto the emergency scene. This can be positive or negative, depending upon the circumstances. We assume that the unstructured time is not breaking any laws or policies and is appropriate for the situation.

  Unstructured time while doing the company's food run would not be as effective/appropriate as unstructured time at the station. This time can give team members a chance to discuss items/activities that are important in their lives—family, sports, cars, and new purchases. Everyone must know that unstructured time is not gossip time. Issues that are important enough to discuss, are important enough to discuss with the appropriate offender or address it with the leader.

If at all possible, the formal leader of the team must be involved in these team formation exercises. An absent leader during formation is an ineffective leader during operations. Think of the Vietnam leadership stories. Sergeants had trained and worked with their teams to the point where they were a cohesive unit. The next thing that happened was the military threw a formal leader, a lieutenant, into the group. Above and beyond the perceived issues with the new lieutenants, such as age, inexperience, and educational levels, the real issue was probably that the new lieutenants had not been with the group during formation and now they were to lead them into very deadly situations. There was no trust between the men and their formal leader.

Smart lieutenants realized that there were formation issues, and that they would have to become part of the group over time and through their actions. Smart lieutenants allowed the sergeants to maintain the main leadership role of the team, which was probably a giant step in gaining the trust of the team. Smart lieutenants gained trust and respect by running interference with the hierarchy and watching out for their teams, and just being in the same situations and experiencing the same dangers. In time, lieutenants learned more about how to lead people, and slowly assumed the leadership duties from the sergeants with permission from the men and not from above.

The old team

Now that your team has been together for a while, it is time to maintain it. Team formation and human interaction is never a destination—it is always an adventure. Your team will have highs and lows during its life. It is the leader's job to reduce *how low* the lows are and increase *how long* the highs last. Think of the leader as the person who has his/her finger on the pulse of the team constantly. That leader must always detect that one missed heartbeat or that one irregularity that will decrease the effectiveness and safety of the team. Throughout CRM you will learn interrelated skills that will assist you in detecting, diagnosing, and repairing problems within your team. These problems include personal readiness of your team members, communication issues, and hazardous attitudes. All these skills of recognition and repair can keep your team hitting on all eight cylinders.

Skills to maintain a team are similar to those of building a team, and there are some additional skills that you will need to become proficient at keeping your team together and functioning correctly. It is not unusual for teams to perform below the standard when forming, and then surpass the standard, possibly by a great deal, and then fall back to a standard or marginally standard performance level.

I think we can all attribute the initial poor performance to rather simple factors such as inexperience with the new team members, unknown criteria for success, personnel issues, leadership issues, etc. The surprising part for team performance is not that they have substandard performance early on, or even that they excel after being together for a short time. The surprise is that the level of performance sometimes drops back to minimal levels after a period of time. The attributes of falling performance in long-term teams seems to fall into a few categories.

1. **Leadership Issues.** Lack of a present, formal, effective, central leader seems to be a major factor in failing teams. Leaders need not control all aspects of the team or make every decision. But leaders must ensure issues that need to be addressed are addressed, decisions that need to be made are made, and factors that affect the performance of the team are identified and dealt with appropriately.

    Leaders who are devoting too little time to team maintenance or are more interested in a particular area, will see the team's focus also shift, sometimes with detrimental results. Leaders must not trade long-term success for success in the short-term. In short, smiles today do not equal safe, effective operations tomorrow—or a strong team.

2. **Personal Issues.** People have problems in their lives. The standard speech from the organization and its leaders sounds like, "Leave your personal problems at the door." Anyone who

is human knows that no matter how hard you try, there are problems that affect how you work, how you think, and how you cooperate within your team. Those problems should not be ignored by the leader. Neither should the leader hang out a shingle over the door and put a couch in the office. Personal problems of the members will affect the team and should not be systematically ignored. They are work-related, because they affect work and the team.

If both your work life and home life are happy, then you are successful and settled. If one or the other is not as happy as it should be, you may still be successful and effective most of the time—one positive carrying the negative created by the other to make you an overall successful person. But when both home and work are unhappy, you are definitely in for trouble all around. Counseling, career and life changes, and other adjustments may be the only way to salvage yourself, your family, and your team.

The leader should be proactive in addressing these issues. There are various methods used today, and even more that are waiting to be discovered and implemented. Some current examples are employee assistance programs, psychological training and testing, medical insurance for psychological issues, and others. Some possible opportunities for leaders to succeed, also fall into the areas of benefit adjustments, scheduling changes, overtime requirements, job sharing, and cross training, etc.

3. **Organizational Support.** Like the example above with the new training program and the lack of fitness funding, organizational support is crucial to team effectiveness. Good leaders know how to point to a job, provide the resources, and then get the heck out of the way. On the other hand, effective organizations know that their people need support and guidance to be successful. Too many times organizations think they are helping their members be successful by merely not getting in the way. This lack of support is perceived by the

organization as, "letting them take initiative to make themselves successful," when in reality it is setting the team up for failure. The lack of organizational support, guidance, and control, either before, during, or after the outcome can initially be met with success, but in the long run it is doomed to failure and can destroy a team.

4. **Self-Factors.** Even adults can be compared to a lump of clay. Leaders can mold them into about whatever they need (sometimes this is good and sometimes it is destructive). But like clay, it does not take much to reshape them or completely destroy them. Some firefighters are more resilient than others. Some team members mold and crush easily, while others seem to be made of magic clay and they return to their original shape without any help. How firefighters operate from a position of inner strength will affect how successful they are and how successful the team is.

The extemporaneous teams

Short-term teams are a challenge to lead. Luckily, most of these types of teams have an extensive support system behind them to increase their potential for teamwork, success, and safety. Examples of these types of teams are all over the emergency services. Volunteer fire departments whose members respond as needed may have a choice of 50 teammates. Some ambulance services schedule their personnel as single units, rather than the crew as a unit; they only come together during training or emergency incidents. This means each shift brings a new partner and new issues for teamwork.

Full-time crews may form a strong team but they are often assigned to work with other teams on larger incidents, such as wildland crews who are placed within an incident command structure. This situation requires them to work closely and support or get support from other—totally unknown—teams of firefighters. All of these situations need special attention to be safe and successful.

Sociologist Jon Driessen found, "the greater the crew cohesion, the fewer the accidents."[7] One of the more dangerous times during a wildland fire incident (and if studied would probably be true in structural firefighting also) is a period called transition. This is the point where it has become too large for "local" resources or it has lasted too long for those resources. A transition to a larger, more extended attack type of operation brings with it changes in strategies, tactics, resource availabilities, and command structures. "Two recent studies have found that when both the fire and the firefighting organization are in transition, fire crews are at maximum risk."[8]

When a team is together for a period of time, the members begin to understand, almost intuitively, what is normal for their fellow members. When one member seems a little *off*, it is easy to sense, address, and/or solve the problems. These teams know the strengths and weaknesses of the individual members, can compensate for those weaknesses, and can be compensated for their own weaknesses. Also if one team member is not comfortable with the way he or she is feeling either physically or mentally, it is easier to bring those issues out within the context of longer-term teams.

In an extemporaneous team, few of these luxuries exist. Crew members sometimes rush to get to know each other, and to discover what their expectations are, what tasks they will perform, what they can and cannot do, and a million other mission critical items. Leaders must strive to create the best team atmosphere in the shortest time. Communication is the key in these types of situations. If the systems of communication are not established prior to the team being thrown together, the team will likely fail when presented with nonstandard or unpredictable situations.

Leaders will set the stage for communications from the start. Even while in the apparatus responding to an incident, leaders can communicate their expectations for the individual team members and the team as a whole. By being open about the potential problems, and the expectations of working together and communicating throughout the life of the team, the leader establishes a norm that will bring the team closer to being a true team.

On a large wildland fire, I was in charge of a strike team of engines (six "pump and roll" one-ton units) that were alternately protecting structures, mopping up (overhaul), and attacking the rapidly moving fire. The individuals on these units had varying experience and qualification levels, and all were seasonal employees of the forest service familiar with the type of terrain, fuels, and fire behavior we could expect.

I had made extra efforts in my fledgling CRM manner to address the responsibilities of the leader (me) and the followers (them). I had drilled them on what I expected from them in their communications (questioning not abdicating), their information flow requirements (two-way and proactive, don't wait till I remember to ask for it), when to advocate their position, and how to share the information they gathered in their areas.

I thought, from the day-to-day activities that we had all become quite the little team. I was very proud of our accomplishments. I was soon to find out that the frosting on the cake covered a hideously undercooked surprise. What had passed for CRM and teamwork under non-stressful, mostly mundane, activities was not to hold up under the intensity that was to come (as is the case on most large wildland fires—10 days of sheer boredom surrounded by 10 minutes of sheer terror).

We had been preparing a couple of structures and our plans for protection for a couple of days. We knew that the fire was eventually going to hit our newly adopted homestead—we just did not know when. One hot, windy day it happened.

In the middle of the afternoon, with plenty of burning hours left, the fire advanced on our progeny. The fire struck viciously, but was repelled by some valiant structural firefighters working in our division. Our problem was now not the structures themselves, but the other 20 structures that lay sprinkled up the opposite hillside. The trees were thick and the vegetation overgrown and dry. The roads were narrow, long, winding, with no turn-arounds for miles. The houses themselves were carved into the hillsides and built around the trees. It was a firefighter's worst nightmare. If fire got started on the other side of our control line, we would lose every one of the homes. There was no way the team could go on that hillside.

When the fire hit our control lines, we got spot fires across our line. These are small embers that fall out of the sky and give firefighters headaches as well as backaches. When the first spot fires were reported, I quickly ordered two units to attack! Attack! The firefighters, being uncomfortable with the prospect of this assignment, did just what they should have done—they questioned my tactics of direct attack on a fire in thick, dry fuels, with high winds, in a no man's land.

I had concentrated on the team and our CRM principles. I should have concentrated a little more on what would happen, and what we would do when the ship hit the sand. Luckily I was able to pull out some "background" techniques to accomplish our mission and assure the team. I told them to go up there, I would be there to meet them, and we would hit it together (personal involvement with the troops, works every time). I had made some errors in my training and planning that could have cost us and others a lot. My CRM stance was quickly adjusted for the new information and experiences. Train for the everyday every day, but train for the unexpected constantly.

Because we had all come from the same general area, I assumed that we all had the same general view of the events that would unfold when our fire eventually decided to show up. I did not do a good enough job preparing for what would happen, discussing what we would do, and brainstorming the possible problems we all expected and how we would/could handle them.

Detailed pre-briefings, coaching during the downtime, and more status reports and size-ups would have better prepared us for this incident. But CRM worked perfectly. The crews did not feel comfortable with the assignment, and wanted more information before committing to something we had all been warning them about. Then, by inquiring about the assignment, I was reminded that I needed to heighten my involvement to improve safety and control of the operation.

PRACTICAL LEADERSHIP

There comes a time when, after all your hard work, training, and preparation, you must actually perform under the stress of the incident. Emergency incidents can usually only go one of two ways: the standard incident or the nonstandard incident. Not to downplay the urgency and danger of our incidents, but most of the time our jobs are what we expect. Not much is surprising on these standard incidents and they are handled "by the book." Nonstandard types of incidents fall outside of our normal training, preparation, and thought processes. Pre-plans, SOPs, and policies can only go so far in preparing the firefighters and the leaders for these incidents.

Responders must use all their skills, experience, and knowledge as well as those of their team members, and other resources, to safely and effectively mitigate these nonstandard incidents. We must not fall into the trap of applying conventional wisdom to unconventional problems. When faced with unconventional situations, sometimes unconventional tactics must be employed. Good examples are found every day across the nation. Some of the most famous are tactics used to combat the oilfield fires in Kuwait after the first Gulf War.

The firefighters who were tasked with extinguishing hundreds of dangerous oil well fires over a period of years, decided that their standard method for combating and plugging the wells was going to take too long, be too dangerous, and require too many resources. They used unconventional tools to accomplish their task in a much shorter time with increased safety. Some of the tactics used had never been used on oil well fires. Jet engines combined with water and chemical mists were just one of the unique tactics that proved effective. By using the jet engine like a giant blower, the firefighters were able to combine water streams and a cloud of fire-extinguishing powder, and apply them to the fire from a safe distance. This technique proved very quick, easy, and reduced the resource requirements of the fire teams.

It is easy to fall into the trap of the standard incident, by looking at all calls as if they were standard. "We've responded to this type of call a thousand times," is often overheard in the apparatus on the way to the call. Sure, you have been on a thousand and one dumpster fires (or grass fires, or difficulty breathing calls), but just maybe the thousand and second call is the one that is nonstandard.

There are few rules that fire plays by—or humans for that matter. As a leader, you must always guard against you or your team becoming complacent. Maybe this dumpster fire is the one that contains the discarded meth-lab; or the grass fire is the one started just before the cold front approaches; or your truck breaks down during direct attack; or the difficulty breathing call is a household hazardous materials scene or the scene of a violent murder. It is the leader's job to always ask the team, "What if," to prepare the team for this, or the next, nonstandard call.

What if's

What if is an excellent tool for combating complacency, providing micro-training opportunities, warding off boredom, and mentoring potential leaders. Asking "what if" helps to identify safety issues. Asking "what if" reinforces standard training and SOPs, such as protective clothing needs, safety procedures, attack strategies, etc. Asking "what if" transfers expectations of the leader (and the organization) to the team. Asking "what if" takes the standard training, expectations, and thought processes into the realm of the abnormal incident. It is difficult to train for abnormal incidents, but engaging in "what ifs" can allow the team to brainstorm solutions for those unusual situations—"Thinking outside the box." These opportunities are often overlooked by the leader and the team.

Complacency is the norm rather than preparation for unusual events that may seem unlikely. What is the likelihood that the next car fire will contain a bomb? (What about a trunk full of fertilizer and fuel?) What is the likelihood that the next grass fire will be a hazardous materials scene? (Clandestine drug labs are using our public lands to manufacture drugs and then discard toxic by-products.)

What is the likelihood that a drunk driver will drive through your scene? (Roadways are one of the most dangerous theaters of operation for us.) What is the likelihood that a giant pink bunny will attack one of your firefighters? The point is, we hardly ever address those "what if" situations in a way that prepares us for the situation we are responding to, the situations we might face, and helps to transfer information that may be applicable at a later date. As a leader you must address those situations to allow you to set your team up to succeed in the most difficult of situations. You are preparing your team for, "The Big One."

Technical proficiency

Let us face facts. Leadership is no walk in the park. CRM leadership may be even more demanding than that leadership of days gone by. Why then would we want to adopt it, if it will make a difficult job even more demanding? The answer is, if for no other reason, safety. CRM leadership will make operations more effective by using more information. CRM leadership will increase morale, because the followers are not just pawns that leaders play. Their input matters and has an effect on the operation as a whole.

As we stated before, the situations the modern fire service is faced with are just too demanding in knowledge, training, and experience to expect a single individual to hold all the cards. In a training program, you need to transfer knowledge and skills of the job first and foremost. People must be technically competent to contribute as team members. While being introduced to CRM, a fire department member called this technical proficiency requirement into question.

> He said, "As an officer on a wilderness search and rescue team, I needed everybody I could get." I stated an example in contrast. "So, I'm walking down the street and you grab me, throw me into the unit and tell me that we are going to hike up 2000 vertical feet, in the wilderness, in the winter, and search for a downed airplane. I have no winter or wilderness survival training. I don't know how to utilize the Electronic Locator Beacon (ELT) to find the airplane; and, I'm not in the best

physical shape. What good am I going to do you?" He said, "Well...you could go get groceries or something." I said, "Exactly! I can do that! Grocery shopping I'm technically proficient at doing."

Disproving his own argument, he agreed that technical proficiency in your job function was a requirement. Using people on emergency scenes who are not up to a minimum technical competency level is setting up everyone to fail—maybe fatally.

After you are technically proficient in your job function, you apply CRM to employ that proficiency in a team setting. Technical proficiency without teamwork is like bad arithmetic—1+1+1=2. While even marginal technical proficiencies with teamwork is like *super* arithmetic—1+1+1=5. CRM has a synergistic effect on your operations. As a leader you can use CRM to enable you to do much more with less, while maintaining your safety margins. Without CRM, people do their jobs almost as if in a vacuum, often repeating tasks and overlapping responsibilities due to poor teamwork and communications.

CRM enables you to use micro-training opportunities to increase the knowledge and readiness of your team. We are not implicating that you explain a basic skill on-scene. This is better suited for traditional training methods. What we are implicating is that the scene is a perfect place to transfer little jewels of information that will increase current understanding and effectiveness, and will transfer to other situations and further increase the effectiveness, safety, and understanding of your team.

WHERE THE RUBBER MEETS THE ROAD

So how do we apply all of this information in the real world? You must first know that there are still millions of dollars and thousands of hours of research going into this field. They are discovering "best practices" every year it seems, and we can learn valuable information from these new "discoveries." So what you learn or reinforce here needs to be continued through application, feedback, and further learning.

Leadership practicals before the incident

As we have said earlier, one of the main responsibilities of a leader prior to the incident is to build an effective team. We will not go over this in detail again, but remember that preparing your team is a function of technical proficiency. CRM proficiency includes pro-active planning for expected responses (standard incidents), and unusual responses (nonstandard incidents—what-ifs), and the leader setting the team climate. Success at this stage leads to a more successful, cohesive team during other stages.

Leadership practicals during the call-out

Leadership does not begin and end during the incident or task. Your leadership carries over from day-to-day, person-to-person. You set the climate of teamwork and communication prior to the tones being struck, and you continue through the call, and back at the station (and some may argue that your leadership status is engaged even when off duty). When the call comes in, many times you must make decisions that could affect the first critical minutes (or hours in rural settings) of an incident.

During a call-out to an oil well work-over rig that had tipped over and was trapping a man, a decision made immediately after the radio message made the difference between success and failure (or at least a substantial delay). A junior firefighter suggested that we should take a firefighter off the rescue truck and have him respond in the depart-ment's winch truck (a truck setup for heavy lifting and pulling, basically for the mechanical division and generally not used on emergency scenes).

The incident location was 15 miles from the station, or about 25 minutes, due to poor access roads. When the rescue unit arrived, the firefighter found a man who was in traumatic cardiac arrest underneath the rig derrick. The ground was soft and initial attempts to lift the heavy structure with traditional hydraulic tools proved useless due to limited space and the soft ground.

Immediately after the first attempt to lift the structure, our winch truck arrived on-scene and was placed into service. While Advanced Life Support was being administered, the lift was planned. Using on-scene, non-fire department workers who were familiar with the weights involved, centers-of-gravity, and construction of the rig, we attached the lifting system and quickly raised the structure off the patient.

The lessons learned are

1. Initial quick thinking, use of non-fire department experience, an open, supportive climate, and application of nonstandard procedures assisted the firefighters with their main goal of extricating the patient. This process allowed EMS to provide patient care, and allowed them to do this in a more timely manner.

2. On-scene use of non-emergency workers who were familiar with the rig was paramount in the success of the extrication.

3. Teamwork between the rig workers, EMS, and fire department personnel was effective because of work and training performed prior to the incident, and applied during the incident. Although it was known that other equipment was forthcoming, the actual operation continued using traditional techniques until "Plan B" arrived, thereby not delaying care and techniques that very well might have worked, if circumstances had been just a little different.

En route to the incident

While en route to the incident, there are sometimes missed opportunities for teamwork and leadership. Riding in the unit is not like a ride on a subway with strangers—although the silence sometimes makes it appear that way. Use this "trapped" time to transfer pertinent information and SOPs to the team. Make assignments, discuss the possibilities of a standard or nonstandard incident, and remind everyone of their CRM responsibilities.

Too often the communications of this sort are taken for granted because the team has been together for so long or the incident does not seem critical. Do not delay communications for a time when stress could adversely affect their transmission and receipt. As a leader, you should actively use this time to communicate actions, strategies, and tactics with the team. For those calls that seem standard, ask team members what they might expect to go wrong and what they would do. Ask them to come up with a "Plan B," if things are not as they expected them to be when they arrive on-scene. All these procedures will strengthen the team, and impart leadership traits to other members.

On-scene responsibilities

Some personnel (both junior and senior) view the scene of action as the time and place where leadership takes place. I hope through our discussions, you believe this to be far from the truth. Actually, the scene of action is the place where all your hard work before the incident is put into play. Your technical proficiency and that of your team is put to use. Your CRM discussions, training, and practice are put into use. Also, your ability and your team's ability to think on their feet and communicate that information are tested.

Of course, on-scene leadership is not all about a team making consensus decisions. Leadership involves someone who is responsible for personal actions, those of the team, and the department. Sometimes leaders make decisions that are unpopular or even sometimes ineffective. Stuff happens, as they say. What is important is not entirely dependent on whether the fire goes out or the patient lives (as in our previous example).

Although we are there to perform heroic jobs, we must not lose track of our number one priority, "Everyone Goes Home Safe." Keep this in mind when making assignments and decisions, and your job will suddenly change from muddied flood waters to a crystal clear babbling brook. If all your decisions have to meet that one criteria, it suddenly becomes at least a little easier to make the decision to sacrifice 20% more of the structure to the fire, or add another 100 acres to the wildland fire, or cut another door in that brand new

Mercedes-Benz. Sometimes our decisions are influenced unnecessarily by outside factors regarding our efficiency, public outcry, perceived duties, etc. The principle that must be foremost in our minds is that "Everyone Goes Home Safe."

After you realize what your main priority is on-scene, start making some decisions about what to deploy, where to deploy, etc.; this is really standard fireground leadership stuff covered by a large variety of very good textbooks. As a leader, you must be familiar with all of the firefighter duties as well as those additional items needed for your role. These include things like building construction, advanced fire behavior, and crew capabilities. These are the technical proficiency portions of your job.

The CRM proficiencies for an on-scene leader are very dynamic. There are some definite events that occur, but like the emergency medical technician field, there are also some pertinent negatives that could be the clues you need to make sure your operation is safe and smooth.

Feedback

When the operational strategies are assigned, you should expect to get some feedback from the firefighters. Feedback could include questions that attempt to clarify their exact duties or where they fit into the scope of the entire plan. *Example*, "So we should pop the back door when the vent team gives us the sign, or when you do?" Or they could attempt to understand why they are doing something (this is usually the case in nonstandard incidents and operations). *Example*, "I didn't think we were supposed to spray water in the roof vent hole?" In effective, long-term teams, questions do not always appear—especially on standard incidents with standard assignments. But when situations get stressful and nonstandard events are taking place, the lack of questions is a pertinent negative.

When as a leader you give an assignment that is out of the ordinary or even rare, and you do not get a question or two, you should be worrying about the effectiveness and safety of your team.

Have they gone macho? Have they got *mission-itis?* Why aren't they concerned about my order? As with medical pertinent negatives, they may just be jotted down in memory or they may need to be addressed immediately—your judgment will decide which it should be.

An example comes to mind that addresses this situation. A department had changed their SOP on the cutting of the battery cables on vehicles involved in substantial impacts. They once cut the cables on every car involved in the collision, but this began to cause more problems with items that could have been mitigated without extrication; i.e., roll down power windows, move power seats. So they decided that before cutting the cables they would assess the danger/hazards involved prior to making the cut/no-cut decision. This was communicated in SOP format, and addressed during training sessions. But like all volunteer departments, it was difficult to make a change like this. So some examples of questions and pertinent negatives arose on numerous occasions.

An officer decided not to immediately cut the cables on a vehicle involved in a rollover. The order was given and there were no questions. The firefighters cut the cables. Why? That's the way they had always done it and they did not really hear the order not to cut, because of the stress of the situation and previous experience and training.

A senior firefighter on-scene decided not to cut the cables in a two-vehicle accident. The firefighters immediately (probably because of the lack of formal rank) questioned him openly and somewhat aggressively. The cables were not cut, but questions arose as to the reasons why, which gave him the opportunity to explain why. Although a little tense at first, this gave the officer a micro-training opportunity, and gave the firefighters a chance to discover what the thought process was for the decision—and not that they just thought the leader was nutso.

Right after implementation there was a serious two-vehicle, head-on collision with one patient deceased and two others critically injured. The officer gave the order not to cut the cables immediately. No response. (pertinent negative). He again gave the order not to cut the cables, and added that they should try to move the seats rearward, and then they would probably cut them. Then he got the proper response and not just a head nod.

On-scene communications are some of the most incomplete, undecipherable, confusing, and awkward of all communications. Leaders should take extra pains to communicate with their firefighters, and ensure that their firefighters feel comfortable enough to communicate with them—and know how to communicate properly. There probably aren't a handful of people who have ever been on an emergency scene where communications were not at least a contributing factor to a problem during the operations. Every critique of every emergency incident usually has at least one communications problem that is identified.

After the call

After the call is the time to address the confusion of the incident—on the battlefield it is called the "Fog of War." Not everyone on-scene has the luxury of seeing the *big picture*. Firefighters often do their jobs, but are confused about how the jobs fit into the rest of the operation. A critique after the incident is imperative to clear up confusion, address problems, reinforce successes, and transfer information about strategy, tactics, and decision-making to those involved.

More on critiques is included in the debriefing chapter. As a leader you should make sure that your personnel are well aware of the opportunities that an after-incident critique has both at the present time and for the future. Continue to address those teamwork issues after the call and the critique. Items that affected your emergency operation could eventually destroy your team. Take pains to address those things during station duties, training, and other incidents. Some of these items need not be common knowledge among the team; they can sometimes be accomplished clandestinely. As sneaky as it sounds, building a strong team is your responsibility, and you should attempt to do this in every legal, ethical, and moral way.

CONCLUSION

Leadership is a very personal thing to everyone involved—from the leader to the led. Everyone has the perfect set of rules to follow, but no two are identical. It is up to the leader to loosen up the rules by opening up the playing field to others. In darkness and isolation the human animal suffers and begins to form a warped perception of the situation and position. In the light of CRM human animals thrive because they are enveloped by requests for "what they know," and "how they would do it," and "what do you have to add." Not warm and fuzzy stuff, just involved leadership.

In strict leader-led relationships, those with little or no feedback or micro-training opportunities, the individuals wither and the team is negatively impacted by the dark flowers all around. In CRM people are taught how to communicate their concerns and questions appropriately and what their responsibilities are, not only to the organization and the leader, but also to themselves and their teammates. They are started as seedlings of the fire service and nurtured through constant attention and leadership to grow into the strong tree that may provide protection for the next batch of seedlings. Let's face it, our number one priority is that "Everyone Goes Home Safe." The only way to ensure this goal is achieved, is to make sure that everyone's life that depends upon us can depend on us to be there.

REFERENCES

[1] *South Canyon Fire Investigation Report*, NWCG, Boise, August 17, 1994.

[2] NTSB Safety Study, *A review of flightcrew involved, major accidents of U.S. air carriers, 1978 through 1990*. PB94-917001. NTSB/SS-94/01, 1994.

3. McKinney, Jr., E., J. Barker, Jr., J. Woody, J. D. Garvin, and W. Olson, "Recent Experience and tanker crew performance: The impact of familiarity on crew coordination and performance." USAF Academy, CO. In R. Jensen, and Rakovan, L, *Ninth International Symposium on Aviation Psychology*, 1997: 586.

4. Reprinted from Weick, Karl E., "The Collapse of Sensemaking in Organizations: The Mann Gulch Disaster," *Administrative Science Quarterly*, 38:4, Johnson Graduate School of Management, Cornell University, 1993.

5. Paraphrased from "Valuing Leadership in an Era of Prophets, Politicians, and Pugilists." Major Charles T. Barco, USAF. *http://www.airpower.maxwell.af.mil/airchronicles/apj/apj94/barco.html*

6. Driessen, J., "Crew Cohesion, Wildfire Transition, and Fatalities," USDA Forest Service, Technology and Development Program, Missoula, MT. TE02P16—Fire and Aviation Management Technical Services. February 2002.

7. *Ibid.*

8. *Ibid.*

FOLLOWERSHIP

PERHAPS THE MOST under-trained topic in the fire service arena is *followership*. It is also the most difficult to describe in concrete terms. Phase III of the Wildland Firefighter Safety Study[1] made the following recommendations as to what type of training a wildland firefighter should receive.

Firefighters—one hour of followership as a stand-alone module on

- Interaction

- Listening skills (listening and learning)

- Receiving, interpreting, and following instructions

- Teamwork

- Good followership (attitude, cooperation, influence, leadership, initiative, work ethic)

- Making decisions together

- Watching out for one another

Good followers are more than mindless robots that blindly execute whatever the leader demands. Instead, good followers support their leaders through technical expertise by compensating for other, less capable team members, and as "translators" of the leader's intentions into practical actions by the team.[2] This chapter analyzes those topics as well as others in an attempt to modify the way we think about followers and how to be followers.

WHAT IS FOLLOWERSHIP?

Followership is not blind obedience. On the other hand, it is not insubordination. Followership is the skill of being the best contributor to the team a person can be. In football, 11 people are on the field. There is one quarterback, a combination of running backs, a combination of receivers, and five down linemen. Not every person can be the quarterback. Not every person can be a lineman. Each contributes to the success of the team by operating in a defined role. Each person contributes to another person's success.

Sometimes a lineman might observe that a particular pass rusher has certain tendencies that the offense might capitalize upon. The receiver might notice that the defensive backs react in certain ways.

In mid-play a lineman is not going to change the play that the quarterback has called. However, in the huddle the information will be passed on.

A receiver may signal to the quarterback a change in a pass route, or that they are wide open in mid-play. The reactions are appropriate for the time and circumstance and are a result of the competence each of the player's displays in individual positions to assess the circumstances, and pass the information to the right decision-maker.

A football team is a team effort. A great quarterback is nothing if his receivers cannot catch the football. A great receiver is nothing if the quarterback cannot throw the football. Neither the quarterback nor the receivers can amount to anything if the offensive line does not provide protection. Players know their positions, their roles in the overall scheme, and each has prepared himself physically and mentally for the job ahead.

The fireground is a different application of the same principle. The best fire command officer in the world cannot make the right decisions without the right information. The best firefighter in the world cannot save lives and protect property if the firefighter is given inefficient or unsafe assignments.

Followership then is three things. First, it is the ability to contribute to task and goal accomplishment. Second, it is the possession by the firefighter of the skill set necessary to be technically capable, to understand the environmental cues, and to communicate those cues in a respectful and time-appropriate manner. Third, it is not a challenge to command authority, but neither is it unthinking compliance with directives— especially if those directives might impact the safety of the operation.[3]

TENDENCIES OF JUNIOR PERSONNEL

Followers have some specific tendencies, which if anticipated, the consequences of their actions can be avoided. First, aviation research found that junior first officers usually tend to delay actions, rather than begin action too decisively or too early.[4] The implications of this research for firefighters are that a firefighter will wait until a building collapses before pulling out, rather than pulling out when the signs of the collapse become imminent.

Next, the research showed that inexperienced officers' decision-making processes follow a pattern. They tended to jump to a drastic conclusion early in the decision-making, but then delayed acting upon the observed situation, and observed that the situation did not justify the drastic action.[5] A junior officer may take no action, claiming that the command officer has the situation under control.[6]

As we discuss in the last chapter of this book, it was knowledge of this tendency that allowed us to get buy-in into the CRM program from our senior officers.

Finally, the inexperienced officer will not use the time available to attempt a small corrective action that might have helped keep the situation under control. An old, grizzled division supervisor from the real boondocks of Wyoming understood this tendency. Often one of his divisions would call and say, "We need more drip torches in our division. Can you get us some?" He would say, "Uh-huh. Why don't you give me a while and I'll get back to you." What he really intended to do was nothing, because he knew the division was not trying to solve the problem.

Invariably, the division would call back in 10 minutes or so, and tell him they found another source of drip torches, and that he really did not need to worry about getting the torches. It was this command officer's experience and knowledge of the fireline that led him to be able to assess what requests were real and what requests were spurious. Through his experience he knew when it was important to react, and when it was important to subtly shift the monkey from his back onto the subordinates.

Self-expectations

You are responsible for your health. You are responsible for your safety. You are responsible for your training. Other people can have an effect on all of these issues, but no one can tell you how you feel, whether you are at risk, and what you do not know. You alone bear the responsibility for you.

Before a firefighter responds to a call, there are several questions he should answer for himself. If the answers are unacceptable, he shouldn't go on the call.

Physically fit

The first question is, "Are you healthy and in shape?" Are you rested? Do you have a cold or the flu? Is your nose runny? Do you feel any aches or pains? Is there anything about your physical condition that would prevent you from effectively fighting fire, or worse yet, would hamper the rest of your crew in the firefighting operations? These are commonly asked questions, but a good fire call has the effect of tossing common sense out the window.

While firefighting is a dangerous profession, the majority of firefighter deaths over the last twenty years have not been from tragic incidents where firefighters have been trapped in raging infernos. These deaths stemmed from the more mundane and totally preventable causes— heart attacks and motor vehicle crashes. Of the fireground deaths over the last twenty years, more than 45% were from on-duty heart attacks.[7] Amazingly these death statistics do not carry over to large, federal wildland fires.

The recent adoption of a fitness standard, "The Pack Test," reinforces a cultural norm of physical fitness that exists in the wildland theater. How many structural firefighters would never think of walking five miles up a mountainside to fight a fire for 12 hours, but they willingly jump into structural personal protective equipment and enter a burning building. Physical fitness for structural departments is not yet the norm, as demonstrated by the statistics.

As a fire service, we need to take an active role in ensuring the fitness for duty of the other members of our crew, by enforcing daily, vigorous exercise. When the primary cause of death of the emergency responders is heart attacks, which are often predictable and preventable, we, as a service, need to evaluate our priorities and address known problems.

The second leading cause of deaths is motor vehicle accidents. Twenty percent of fireground deaths during the last 20 years were associated with motor vehicle incidents (13.9% with motor vehicle crashes, and 6.6% for other motor vehicle associated deaths).

In responding to the scene and returning from the scene, firefighters (in fire department apparatus or not) must be cautious. The first advice a chief gave me as a firefighter was that I didn't do anyone any good if I didn't make it to and from the scene.

Sixty percent of firefighter deaths fall into the clearly preventable, non-emergency related category.

Fatigue

The next question a firefighter needs to ask is, am I adequately rested? During the period of 1989–1993, an estimated 56,000 crashes on U.S. highways occurred in which driver drowsiness was cited as a cause of the accident.[8] During the same five-year period, drowsiness/fatigue was cited as a factor in an annual average of 1357 fatal crashes, resulting in 1544 fatalities.[9]

In a recent analysis of a fatal wildland fire, fatigue, sustained wakefulness, or sleeplessness were named as factors. Moderate levels of sustained wakefulness produce performance impairment equivalent to or greater than impairment observed at levels of alcohol intoxication deemed unacceptable when driving, working, or operating dangerous equipment.[10] Being tired may be more dangerous than being awake and drunk. In other words, our thought processes and reaction times are slower and more erratic when we are tired than when we are drunk and awake.

Firefighting is a difficult and strenuous occupation, which requires strong physical exertion over long periods of time. Fatigue is a part of the job, but excessive fatigue and inadequate periods of rest are more dangerous than responding to calls under the influence of alcohol.[11] Fatigue affects judgment, coordination, and ability. No firefighter should respond unless adequately rested and mentally alert.

Nutrition

Since we have all experienced grade school, we have been taught to eat healthy. It goes without saying, that what you put into the human body is what you take out. Instead of eating fruits and berries, however, most folks tend to eat processed flour and sugar, and grease.

Bad diet kills your body from the inside out. It lines your arteries with glue, and it rots your stomach and kidneys. Good nutrition, on the other hand, will increase endurance, provide strength, and improve mentation.

Hydration

Drinking the proper amount of fluids is crucial to performance. Studies have shown that losing as little as 2% body mass due to dehydration can lead to decreased performance capacity and damage to bodily functions.[12] It is important for firefighters to maintain good hydration. It is too late for a firefighter to get hydrated in the engine on the way to the call.

It is important for firefighters to maintain good hydration while on the scene. A good rule of thumb is two bottles of water for every bottle of air, and one bottle of electrolyte replacement drink (i.e., Powerade or Gatorade) for every two to three bottles of water. Maintenance of proper hydration will increase performance and decrease accidents.

Mental attitude

The next question a firefighter should ask is, "Do I have a good mental attitude?" The attitude with which you face a fire can determine your success or failure in a fire situation. To be a good follower, a firefighter should become familiar with individual tendencies of each weakness of each crew member. The FAA completed a study and found that most pilot-induced accidents were caused by crews who had been together for less than a month.

In a study of United States Forest Service field crews (those that work together on a daily basis and have duties that include wildland firefighting), Sociologist Jan Driessen showed that crew cohesion and accident rates worked on an inverse correlative relationship.

Simply put, crews are predisposed to accidents and problems until crew cohesion occurs. He found that good crew cohesion takes effect after approximately six weeks. [13]

In the fire service we are obligated to work with unfamiliar crew members; however, a good knowledge of self will help avoid tendencies to enter extra hazardous situations.

The FAA, in their Pilot Judgment Program, has created a self-evaluation program regarding attitudes of pilots. Those five hazardous attitudes outlined by the FAA for pilots have significant application to the fire service. Attitude in responding to an emergency can mean the difference between life and death.

The five hazardous attitudes as outlined by the FAA are[14]

Anti-authority: "Don't tell me!"

This attitude is found in people who do not like anyone telling them what to do. These folks think, "No one can tell me what to do." We all have the right to question authority, but a person who resents authority for the sake of resenting authority is dangerous to himself and to other members of the team. The anti-authority attitude creates a breakdown in the efficient workings of the team. The attitude breaks the team into pieces before it has ever been given a chance to congeal.

Impulsivity: "Do something—quickly!"

This is the attitude of people who frequently feel the need to do something, anything, immediately. They do not stop to think about what they are about to do; they do not select the best alterative—they do the first thing that comes to mind.

They are the people on scene shouting, "Do something—anything!" As professional emergency service providers, we need to have a plan and to act within our training to mitigate emergencies. Charging ahead without a plan (freelancing), destroys the team approach to the operation, and frequently places the freelancer and the team at greater risk than necessary.

Invulnerability: "It won't happen to me."

Many people feel that accidents happen to others, but never to themselves. They know accidents can happen and they know that anyone can be affected, but they never really believe that they will be the involved. Pilots who think this way are more likely to take chances and run unwise risks, thinking all the time, "It won't happen to me!"

Tom Lubnau, one of the authors of this book, was almost killed by this attitude. He entered a building, which was known to be dangerous. It had been burning for too long, and was literally falling apart at the seams. It collapsed on him (see Fig. 7–1). The arrow indicates where he was pulled out. He was in the building for all the wrong reasons. He knew all the risks, and had all the training, but didn't believe anything could happen to him. He almost paid the ultimate price, for a building that was a total loss before he got there.

It was lying under a pile of debris, short one dimension, that Tom became truly interested in CRM by asking the question, "What on earth am I doing here?"

Fig. 7–1 One result of "It won't happen to me!"

His wife summed up the entire situation when she visited him in the emergency room of the hospital and said, "What were you trying to do, save a couch?" (the couch is in the upper right-hand corner of the photo). She then added dryly, "Your life is worth more than a couch!"

This incident might be considered an isolated stupid incident, but as this is being written, one firefighter from a major city's highly trained career department is lying in a hospital. Another is dead because a wall on a building that was lost collapsed on them. Their lives were also worth more than a couch. We keep killing people for couches, and we have to change how we think to keep from killing and injuring more people for couches.

Macho: "I can do it."

People who are always trying to prove that they are better than anyone else, think, "I can do it!" or worse, "I'm the only one who can do it." They "prove" themselves by taking risks and by trying to impress others. Instead, they risk themselves and the safety of others. While this attitude is thought to be a male characteristic, women are equally susceptible, especially those women who are in a bid to "prove" they are as good or better than "any man." The can-do attitude of many of the elite units in the fire service is an accepted strength. But the "can-do, no matter what" attitude kills firefighters.

Resignation: "What's the use?"

People who think, "What's the use?" do not see themselves as making a great deal of difference in what happens to them. When things go well, they think, "That's good luck." When things go badly, they attribute it to bad luck or believe that someone is "out to get them." They leave the action to others—for better or worse. Sometimes such individuals will even go along with unreasonable requests, just to be a "nice guy," or they perceive their actions as less important than other team members.

In addition to the five FAA hazardous attitudes, retired USAF Officer Tony Kern, one of the great thinkers in the human factors arena, and the director of flight operations for the U.S. Forest Service suggests two more hazardous attitudes.[15]

Pressing ("Get-home-itis")

Critical judgment errors are made in the name of "getting the job done." My grandfather once told me to spend the time doing the job right the first time, rather than doing it over again. I often remember his words, but don't always heed them. A case in point—this last summer I was dispatched to a mutual aid wildland fire in the hills north of Rozet, Wyoming. There was absolutely nothing glamorous about this fire. A patch of ugly ground was set on fire by an electrical spark at an oil well. The fire burned through an old stand of sagebrush. Initial knockdown was quick, easy, and uneventful.

The rest of the evening was spent in mop-up, digging up smoking sagebrush roots and ant piles. After about four hours of mop-up on this ugly little fire, I got tired and wanted to go home. I started taking shortcuts. I decided to skip mop-up next to a blade line. Hungry, tired, and wanting to get home to my wife and family, I called it good. The rest of the story is obvious. The next day when the sun heated things up, the fire rekindled and a crew was back on that same patch of ground, putting out the additional 40 acres of ground that burned, because my mop-up was sloppy. Similar stories have plagued the structural service with sloppy overhaul and then resulting callbacks.

We all find the temptation to want to get the job done. We should listen to my grandfather's advice. If a job is worth doing, it is worth doing right the first time.

The airshow syndrome

Showing off for the crowd is one of the most dangerous attitudes we face. When you are thinking, "It's time to make a name for myself and impress somebody," it is time to go home. Firefighting is a serious business. It is not a business for grandstanding.

A fine example of the airshow syndrome in action is a fire department from the east coast who invited the entire community to see the graduation of their latest class of rookies. They built a demonstration house, set it on fire, and called the rookies to come and put it out. With crowds on hand, balloons flying, and an announcer giving a play-by-play, the rookies pulled up in their fire engine, donned their Personal Protective Equipment (PPE) as trained, charged a line, and entered their burning demonstration house to show what firefighters could do. The demonstration house had flashed over about a minute before the firefighters entered the building. But they were determined to put on a show for the crowd.

The videotape of the incident shows the rookies crawling into a room where they are surrounded by flame, top, bottom, and every side. Then the videotape shows the burned firefighters with PPE on fire and melting, jumping out of the burning house. There is nothing more embarrassing than having the fire on your back extinguished at the demonstration, instead of putting the fire out yourself. If you are thinking about showing off either publicly or professionally, think again. You are losing mission focus. Do the job. Firefighting of itself, is a spectacular activity. Don't wreck it by getting hurt while seeking glory.

While we all have hazardous thoughts from time to time, self-discipline to keep your mind on task is the key. Assess yourself. Avoid the tendency to show off. Doing what we do is showing off enough.

The antidotes

Since you cannot think about two things at once, one way to keep from thinking a hazardous thought is to think another thought. By telling yourself something different from the hazardous thought, you're "taking an antidote" to counteract the hazardous thought. You remove a hazardous thought by substituting the antidote.

Thus if you discover yourself thinking, "It won't happen to me," mentally tell yourself, "That is a hazardous thought." Recognize it, correct it, and then say its antidotes to yourself.

To do this, you must *memorize the antidotes.* Know them so well that they will automatically come to mind when you need them.

Poison	Antidote
Anti-authority	
"Don't tell me."	"There is a reason for this rule."
Impulsivity	
"Do something—quickly!"	"Not so fast. Think first."
Invulnerability	
"It won't happen to me."	"It could happen to me."
Macho	
"I can do it."	"Taking chances is foolish."
Resignation	
"What's the use?"	"I'm not helpless. I can make a difference."
Pressing	
"Let's get this done and go home."	"If a job is worth doing, it is worth doing right the first time.
Airshow Syndrome	
"I am going to look so good."	"I am going to get the job done right."

Care should be taken in responses to insure that the hazardous attitudes do not affect fire performance.

HURRY-UP SYNDROME[16]

How many times has a tone gone off, and in a rush to get on the engine you leave behind your hood, gloves, or helmet? In addition to the hazardous attitudes, there is another psychological phenomenon know as the, "Hurry-Up Syndrome." The more serious the incident, the greater the rush, and the more likely it is that a firefighter will make a hurry-up mistake. When firefighters operate under time pressure, they tend to make mistakes. In much the same way, aviation's worst disaster, the terrible KLM/Pan-Am accident at Tenerife, involving two fully loaded 747s that collided on the runway, was due in great part to schedule pressure problems experienced by both flight crews. After studying 125 incidents in which time was a factor, the investigators concluded that the hurry-up syndrome was real, and that it could be consciously dealt with.

Time is unique in an emergency situation. It can be both accurately measured and subjectively perceived. On the fireground, our subjective view of the time available until some event occurs (flashover, frontal passage, building collapse, victim death) and our perception of time progression directly affect our decisions and operations. As firefighters, we tend to operate on a perceived time deficit pressure. Actually if objectively studied, we would probably find about 60–80% or more of our "snap decisions," those made without adequate planning, are unnecessarily rushed.

Very rarely are we faced with a blowup disaster like Storm King Mountain in 1994, which threatens our lives and our senses. When the situation happens, we can fall apart if our habits are not good CRM habits. In that fire it became apparent that the decisions—to run or not, to carry tools and packs or not, to go right or left, or to run into the drainage or not—were decisions that could not be made with the luxury of time. Mere seconds separated those who lived and those who did not.

A great possibility exists that a time will come when the decision has to be made immediately. Try this mental exercise. At your next training session, try to imagine a situation where you and your team must rely on a snap decision or order. If we were on a vehicle incident, and a firefighter yells, "Run," would your team drop everything and run away? Would there be a split second where you doubt what is happening, the validity of the decision, or where you just look up in amazement? Could that cost you? What can you do to prevent this from happening?

One of the precursor fires to Storm King Mountain was the Mann Gulch Fire. Norman MacLean detailed the fire in his book, *Young Men and Fire*. A crew was being chased up a gulch by a blowup. Fire was nipping at the crew's heels. The leader, Wag Dodge, started another fire, and jumped into the black that was created. He ordered his crew to do the same. There was no time to explain the order. It was time to act and live, or not act and die.

The crew looked at him as if he were a crazy man. If they didn't have enough fire to deal with, here the command officer was lighting more fires. Dodge hid in the black safety zone created by his escape fire. His crew, not understanding what was happening, ignored his order and ran up the slope. All of the crew died, except Dodge and two others.

We don't train for handling time pressures and making split decisions under extreme pressure. We should.

Human errors may be categorized as errors of commission or omission. Errors of commission are those in which people carried out some element of their required tasks incorrectly, or executed a task that was not required, which produced an unexpected and un-desirable result. Errors of omission are those in which the person neglected to carry out some element of a required task.

The study revealed that 60% of the mistakes we make when we are in a hurry are errors of commission. Forty percent of the mistakes we make occur when we forget to do something—errors of omission. Time pressures or high workload per unit time were the causes of the mistakes.

RECOMMENDATIONS

To avoid hurry-up syndrome errors, firefighters should consider providing greater structure to firefighting activities in order to reduce the frequency of time-related errors. Similarly when distraction and emergency pressures are seen to occur, a reasonable response is to slow down and carry out tasks in as linear a fashion as practical. Where time-related pressure is encountered from external sources, firefighters may find it a good strategy to calmly explain the nature, probability, and typical results of hurry-up errors to those who "apply the pressure." Additional strategies include:

- Maintain an awareness of the potential for the hurry-up syndrome in response and operational phases.

- When pressures to hurry-up do occur, particularly in the response phase, it is a useful strategy for firefighters to take time to prioritize their tasks. When pressured to make a decision, a firefighter should always ask, "How much time do I really have to make this decision?"

- If a procedure is interrupted for any reason, returning to the beginning of that task and starting again will significantly reduce the opportunity for error.

- Practicing positive CRM techniques will eliminate many errors—effective crew coordination in "rushed" situations will catch many potential problems. Don't allow time pressure to destroy your team.

- Adherence to SOPs and tactical and strategic plans are a key element of response and operational task execution.

- Defer paperwork and nonessential tasks to low workload operational phases.

By following the previous suggestions, many hurry-up syndrome errors can be avoided, thus reducing potentially disastrous results.

Emotional attitude

The professional firefighter understands that having emotions are part of being human. Being in the emergency services, we all have a head full of memories that we would just as soon not have. If critical incident stress is taking a toll on your life, seek help. It is available.

Traumatic stress can manifest itself as depression, grief, anger, guilt, apathy, fear, burnout, or in the worst case, suicide. If such behaviors are observed in a fellow firefighter, it is time to question and perhaps meddle. An intervention is appropriate, just like an intervention is appropriate when someone is in ventricular fibrillation.

Traumatic stress is not the only stressor that can affect performance on the fireground. Home stresses can also influence work performance.[17] While we all have stresses in life, it is the duty of the professional firefighter to assess, deal with, and control the stress and the reaction to stress.

The role of the follower

Much has been written on the qualities of a good leader. Little has been written on the qualities of a good follower.[18] The success of a leader is enmeshed with the success of the follower and vice versa. A leader cannot succeed without capable and intelligent followers any more than a follower can succeed without a capable and visionary leader.

A powerful leader creates success and personal fulfillment in the followers, and the capable and intelligent followers allow the leader to succeed.

In the firefighting environment, a good leader is one who insures all of his followers have the appropriate training and equipment, who understands the task at hand, the strategies for completing the task, and who can motivate the followers to accomplish the tasks. A good leader thrives on the success of his followers, knowing that their success is not a threat to the leader's success, but is a credit to leadership.

In a successful teamwork climate, a good leader will solicit the input of the followers. The leader possesses an open demeanor, some humor, and facilitates the lines of communication between a junior member of the organization and the leadership.

Much has been written about the hesitancy of a junior member to speak up about an incident. Warren Bennis, the great student and teacher of leadership, reports that 70% of followers will not question a leader's point-of-view, even when they believe the leader is about to make a serious mistake.[19] Many examples of this phenomenon occur from simple to complex. A famous example is the meetings between John F. Kennedy and his advisors when planning the Bay of Pigs invasion. No one pointed out the weaknesses of the plan, even though most had serious reservations and issues; a fiasco was the result. A leader needs to know that the followers are willing to challenge his decisions, and must facilitate a proper exchange of information, knowing that other points-of-view will aid in the organization's success.

Peter Drucker, one of the fathers of modern business management, maintains that if a project is unanimously approved by a committee, the project should be sent back to the committee for further study. If the committee has not found any negative aspects of the project, and no dissent has been created, the committee has not fully analyzed the project and its effect.

History is full of examples of nations being lead down the wrong path by uncritical and unquestioning followers. On the other hand, history is full of successes based upon firm convictions, debate, and constructive criticism.

So a successful leader must give permission to followers to provide input and corrections to a plan. Chief Alan Brunacini is a fine example of a leader who is willing, when appropriate, to accept the suggestions of his followers and to insure their success. He strips rank out of discussions by wearing Hawaiian shirts. He tells his people if it is the right thing to do, go ahead and do it; then he rewards people for their good judgment and initiative. He gives his followers an environment in which to succeed in their own right. As a result, the Phoenix Fire Department has become one of the premier fire departments in the world.

On the other hand, a good follower has the following responsibilities. A good follower is responsible for his own training. A good follower is responsible for his or her own safety and the general safety of the organization. A good follower must know the goals and purpose of the organization, and a good follower must challenge a leader, when the leader's actions are contrary to the goals of the organization.

Training

A good follower knows what he *knows* and what he *does not know*. He asks for help and seeks further education at every opportunity. The good follower knows that versatility and knowledge make him a better cog in the firefighting machine. An example that has become commonplace these days, but was not so common years ago, is the firefighter with extensive medical training. What was once a luxury on the fire scene, has now become mandatory. In addition to knowing fire behavior and suppression tactics, firefighters make the emergency services machine run more smoothly also by knowing how to administer pre-hospital emergency care. As a result, the fire service delivers better customer service, and the organization has better success.

A follower is the only person who knows what that follower does not know. It is incumbent upon the follower to let the leaders know what type of training is necessary to insure that good service is delivered to the customer in a safe and efficient manner.

Safety

Although the Incident Command System designates a position of safety officer on the organizational chart, every person, whether leading, following, or just getting in the way, is responsible for the safety of every other person on the scene. While lawyers and juries seek to assess fault to leaders on a fire scene for negligence and harm, the truth of the matter is that each follower is as responsible for the safety of an operation as the leader is. Each person has a role in the organizational safety. It is a failure of responsibility not to act when the risks are acceptable, and the purposes and values of the group would have us act. It is a failure of responsibility to act when the risks are unacceptable and acting endangers the organization's purposes or violates its purposes.[20]

That is not to say that a follower is a leader, or that good followership entails outright insubordination. A good follower knows his role in the organization. A good follower knows what that role is. A good follower also knows when it is appropriate to discuss a matter with a leader. If there is an immediate and articulable danger to life and safety, act immediately. If there is a potential danger in the very near future, point it out. If there was a near miss from which the organization can learn, point it out in the after-incident review.

The good follower knows the following adage, "Advocate your point, but raise your hand first."

Cultural norms

The good follower has the ability to establish the informal norms of the organization. The good follower will only look up to and emulate the well trained, fit, and experienced firefighters and leaders, shunning those who are known for unacceptable risks, bullying, and intimidation. The follower is responsible for the culture of an organization. An example is the reversal of the old adage, that the fire service is 200 years of tradition unimpeded by progress. Actually, during the last few years, the structural fire service has seen remarkable change. What was once a macho culture of "smoke eaters" is generally now considered foolish and unprofessional. The norm is proper protective equipment, breathing apparatus, and a quality retirement.

The wildland fire community is working on such a cultural change, but as with any cultural change, the cultural norms will have to be set and changed by the followers, under the guidance of good leadership.

The follower recognizes there is pressure on the leadership and the "elite" firefighters to put the fire out. Sometimes that outside pressure to put the fire out sacrifices safety. The follower recognizes that pressure, and works to refocus the group effort to professionalism, risk assessment, and risk control as the firefighting culture.

The follower also recognizes that chronic complaining and criticism is toxic to an organization. While constructive criticism coupled with a willingness to solve the problem, is an asset, whining and bickering are destructive to morale. The follower is obligated to redirect the negative and destructive comments into positive and creative energy.

Constructive criticism

There is an old adage that says, "Nothing fails like success, because you don't learn anything from it." I have spent a lifetime learning.

Dr. Norman Vincent Peale said, "The trouble with most of us is that we would rather be ruined by praise than saved by criticism." The follower must overcome the reticence to hear constructive criticism and encourage honest feedback. It is only with that feedback that the follower learns from mistakes. An uncorrected mistake becomes an uncorrected bad habit.

The follower should encourage honest evaluation with statements like, "Thank you. That will help me to be better." Or, "I appreciate your willingness to help me become better."

It takes courage to criticize. It takes more courage to accept criticism and change poor behavior.

Helping the leader lead

We thought long and hard about putting this section into the manual. One of the main criticisms of this program is that we are teaching widespread insubordination. Our goal is the absolute opposite. We would like to help create teams with honest information flow from all levels to other levels. After careful evaluation, we decided to include this section, because to omit it would be to ignore the rhinoceros in the fire service's living room. Situations arise in every department in which poor leaders are elevated to positions of power. Omitting strategies on how to help poor leaders succeed would be the equivalent of ignoring that rhinoceros in the living room, and allowing it to poop all over the carpet. Suddenly the organization suffocates in rhinoceros poop.

Leaders are under immense pressures—to succeed, to maintain crew safety, to get the job done right, to overcome fatigue, and to get the job done cost effectively. The follower is aware of these pressures and works to lighten the load of the leader. Clear knowledge of the organization's goals and values, as well as an understanding of the organization's standard operating procedures, will assist the follower in helping the leader to lead.

Leaders are given little training on how to relieve these stresses and mentally refresh themselves. We expect our leaders to be superhuman paragons of virtue when, in actuality, our leaders are people.

From time to time, it may become necessary for a follower to observe the leader, and assist the leader in finding the way to the organization's goals.

A leader is a human being and from time to time may become overwhelmed at an emergency scene. The tendency for a human being who is in the process of being overwhelmed at a scene is to revert to prior over learned behaviors.

At a fire a few years ago, a young captain was in command. His training was as a hard-core wildland firefighter. He was in command of a fire where everything was going wrong. One of his units had broken down. An old dead tree had fallen over the only way of escape, and the fire was burning down the hill to his position. Instead of remaining calm and in command of the situation, he reverted to his prior over learned behavior, took the chainsaw, and started cutting trees. How often has a command officer been seen on the end of a hose line instead of in command of an incident? When a leader gets overwhelmed by the situation, the good follower will suggest that the leader take a step back and regain situational awareness.

An arrogant leader is toxic to an organization. While on the surface it appears to be a strength, in actuality, communication is stifled, creativity hampered, and initiative dampened. A leader should be confident, but not arrogant. Arrogance should be challenged.[21]

One of the greatest "go to" people in the history of the United States was an arrogant civil war hero and great general. When there was a difficult task ahead, he was the guy who was called upon by command to get the job done, and most of the time he did a great job. He was put in charge of many difficult campaigns during the Civil War, and he succeeded. His superiors did not necessarily like his methods, but they knew he would get the job done, so he became the "go to" guy.

One hot sunny day in June 1876, he came upon a situation he had never seen before. Without good intelligence, he attacked the largest gathering ever of Sioux and Cheyenne Indians in one place at one time. He charged in headlong, without regard for the safety of his officers. On that day, on a hill in southern Montana, George Armstrong Custer, along with all of his troops, died.[22] He died because he did not take the time to assess the strength of his enemy, and to understand the problem he was facing. A classic violation of the centuries old Sun Tzu teachings, "Know thine enemy."

One has to wonder if the Battle of Little Bighorn would have turned out differently, if the general had the capacity to solicit input from his troops.

As with any organization, there are struggles for power. Some leaders will work to satisfy their own ego needs, rather than working for the success of the organization. Power struggles are inevitable, and rarely productive. Ira Chalef refers to an old African proverb, "When elephants fight, it is the grass that suffers."[23] When fire department leaders struggle, it is the firefighters who suffer. While conflict is inevitable, the careful follower will be alert not to get crushed.

An abusive leader is toxic to an organization. Unconstructive criticism, public berating and ridicule are contrary to the spirit of a constructive organization. The ridicule will steal the spirit, the loyalty, and the dedication of the follower, and will gut the organization's morale. Tolerated abuse of any member of the team is unacceptable to the success of the organization. The follower should meet with the leader in a non-public forum and challenge the leader on such behavior.

A follower should never "cover up" the behavior of the leader. Like in the case of an alcoholic, covering up the bad behavior and shielding the leader from the consequences facilitates the bad behavior and allows it to be perpetuated. We should all be responsible for the consequences of our own actions or inaction. Allowing bad behavior to continue is an endorsement of the behavior, and silence in the face of abuse is tantamount to committing the abuse yourself.

How should a leader be confronted?

Given that the follower has obligations to protect an organization, a leader should be confronted in the same way that marriage counselors say spouses should be confronted. The leader should be told

1. What behavior is causing the problem?

2. What effects is the behavior having?

3. What are the consequences of the behavior in as specific terms as possible?[24]

For example, there is an engine company officer who goes to bed before the engine is put back into service. A possible confrontation might sound like the following.

"Lieutenant, when you go to bed before the engine is back in service, other people on the crew get the impression that putting the engine back into service is not important. That type of behavior also affects crew morale. The consequences of your not participating fully in the station duties are that the equipment is not properly maintained and people don't respect your behavior, which causes a breakdown in our crew."

A good leader should accept the constructive criticism in the same way that I suggest the follower should accept the criticism. Constructive criticism, made with the good of the leader and the organization in mind, is a gift.

Too much of a good thing

Now that I've suggested a follower has some obligations to provide feedback to the leader, my advice is to exercise that right with deliberation and discretion. Challenging a leader on specific items from time to time is a healthy practice, but a pattern of constantly challenging the leader all of the time is insubordination. Take oxygen as an example. A certain amount of oxygen is necessary for life. Too much oxygen is toxic. In the same way, a little constructive criticism of the leader is healthy; constant criticism is toxic. The fire service has a four-word label for such behavior. The first three words are "pain in the."

A disgruntled, insubordinate follower will never develop trust of the leader, and is destructive to the operations of the organization. Such a person is a liability to the organization. While dissent and criticism to an organization is important, whining and complaining is not.

After teaching this course, I had a fire chief pull me aside. He said, "I understand what you are trying to teach here. I understand the philosophy and I agree. But I have a problem."

I replied with concern, "What is the problem, Chief?"

"Since you have taught this course, everyone thinks he is fire chief. I had a firefighter call me and tell me he had a doctor's appointment and would be late for work. Four times during the course of the day, different firefighters pulled me aside and told me the firefighter showed up late for work, and that I should do something about it."

I replied, "Chief, that's not what I'm teaching. If you are stepping way out of line, you should be challenged. If you show up on scene drunk, I'm going to say something. It directly affects my safety. I am not teaching people that because they have taken this course, they are all fire chiefs. I am teaching folks that they have an obligation to communicate facts to command. Command can make use of whatever they want with those facts." When presented with the facts, the chief makes the decision. It is then the follower's obligation to follow.

The firefighters' complaining about their fellow employee not showing up for shift, were trying to usurp the authority of the chief. They were trying to force management decisions that were clearly not theirs to make. CRM does not mean that every firefighter has a say in every decision. It means that information should be exchanged to accomplish a task, taking into consideration all of the tools and skills of the team.

The bottom line is don't use this course for insubordination. You should work for the clear purposes of the organization, but respect authority. The chief is the chief for a reason. The firefighter is a firefighter for a reason. Respect those reasons unless someone is going to die or get hurt when the decision is made, the follower follows—period.

Skills to be a good follower

Then, what are the skills that a firefighter needs to develop to become a good follower? Followership is more than just an attitude; it is a skill that can be developed just like any other skill.

Respect authority

The first and most important skill the firefighter needs to develop is a healthy respect for authority. Each person has a role that needs to be filled. Ultimately the role of the follower is to follow. Following can only be accomplished with a healthy respect for the person who is in charge.

To cultivate respect, one must respect both the person and the position. The person who is filling the job of command, in some fashion earned that spot. Respect the knowledge and abilities, which caused that person to earn the spot. One must also respect the position. Although a person filling a command slot may not be the best command officer in the department, the person is filing a slot that needs to be filled. The person in command of an incident has the opportunity to have more knowledge and more insight into what is happening on the fire scene. That person has the global picture. A good follower respects the global picture of the command officer.

Safety eyes

The second skill the good follower has is a responsibility for personal safety and an eye toward the safety of others. Even though we appoint a safety officer on major incidents, the truth of the matter is that everyone on the fire scene is responsible for the safety of every other person on the fire scene. The good follower has a grasp on his or her own knowledge and capabilities and the knowledge and capabilities of every other person on the crew. Knowing what one does not know is as important as knowing what one does know. The follower places the safety of the public and the crew above all other priorities and conducts operations at all times with that concern in mind. Know what the risks are. Risk little to save a little. Risk a lot to save a lot.

Coupled with the responsibility for safety is the authority to make operations safe. Firefighting is a dangerous profession. It will always be dangerous. There will be controlled risks taken at every turn in the road. However, if an unreasonably dangerous operation is about to be conducted, those people with the responsibility for safety (everyone) must be given the authority to advocate for a safer operation, even if it means stopping an operation briefly before it is commenced. With this authority comes great responsibility. If an operation is stopped, it better be stopped for a very good reason, and not out of ignorance. If one undertakes the mantle of authority, that person must have the skills and knowledge necessary to exercise that authority.

Technical competence

Thus the third skill the good follower has is technical competence. The good follower is one who, when asked to perform a task, meets or exceeds the expectations of command in a safe and efficient manner. Technical competence is the place where we spend the bulk of our time training, but it is only part of the skill necessary to be a member of the team. Just like we must practice the skills leading to technical competence, we must also evaluate and understand the skills necessary to be a good follower.

We must not only understand what we have to do, but why we have to do it the way we do it. We must understand how our role fits into the big picture. We must understand the consequences of failure to fulfill our responsibilities. We must know, for example, that if we don't get the ventilation done properly, the entry crew gets blown out the front door by a backdraft. If something prevents proper ventilation, the right people need to be informed, before a bad thing happens.

Communications, again

The fourth skill necessary to be a good follower is good communications. We've devoted an entire chapter to this topic, because it is central to all of what CRM is about. While we've devoted an entire chapter to this topic, it still never hurts to reinforce an important message. We've all seen the dramatic video of the Washington, D.C. rescuer diving into the Potomac River to rescue people who were drowning after Air Florida Flight 90 ran off the end of the runway and into the frozen river. What we don't get to see is the unclear communication that occurred in the cockpit prior to the takeoff.

The weather was bad. The aircraft had been deiced, but 45 minutes had elapsed before Flight 90 was cleared for takeoff. While waiting for clearance, the following conversation occurred:

FIRST OFFICER: *Look how the ice is just hanging on his, ah, back, back there, see that?…*

FIRST OFFICER: *See all those icicles on the back there and everything?*

CAPTAIN: *Yeah.*

The first officer was uncomfortable with the icing conditions. He was hinting to the captain, trying to tell him something was wrong. What was being communicated to the captain?

After a long wait following deicing, the first officer, again, expressed his concern about the condition of the airplane.

FIRST OFFICER: *Boy, this is a, this is a losing battle here on trying to de-ice those things, it (gives) you a false feeling of security, that's all it does.*

What was the captain hearing at that point? What was the message the first officer was trying to convey?

Shortly, after being given clearance to take off, the first officer again expressed his concern.

FIRST OFFICER: *Let's check those tops again since we've been sitting here a while.*

CAPTAIN: *I think we get to go here in a minute.*

Finally, the airplane had begun to take off. The first officer must have noticed something was wrong with the readings on his gauges.

FIRST OFFICER: *God, look at that thing. That don't seem right, does it? Uh, that's not right.*

CAPTAIN: *Yes it is, there's eighty.*

FIRST OFFICER: *Naw, I don't think that's right. Ah, maybe it is.*

Then, a few moments later, they had this conversation.

CAPTAIN: *Forward, forward, easy. We only want five hundred.*

CAPTAIN: *Come on forward…forward, just barely climb.*

CAPTAIN: *Stalling, we're falling!*

First Officer: *Larry, we're going down, Larry....*

Captain: *I know it.*

[Sound of impact][25]

One can imagine the first officer's ghost, sitting in the cockpit, looking at the captain and saying, "You moron, I told you four times we had a problem." Then the captain's ghost looking back and saying, "If you are going to say it, say it directly. Don't sit in the officer's seat and whine."[26] Instead we have 74 of 79 people on the airplane dead and some amazing videotape of a really cold rescue.

The same kind of inarticulate transmissions occur daily on the fire scene. Fortunately for those involved, the International Association of Firefighters (IAFF), International Association of Fire Chiefs (IAFC), or United States Fire Administration (USFA) does not release those transcripts to the world. Unfortunately for the rest of us, we don't learn from the communications mistakes of others, because the mistakes are buried in the ether. So we protect egos, and don't teach each other from our mistakes.

The point—if there is a problem, express the facts and state your opinion clearly and precisely.

Learning attitude

The next skill necessary to be a good follower is to develop a learning attitude. Try to see things through the eyes of a child. Whenever there is time ask, "Why does it happen that way?" Remember the old story about the daughter who always cut both ends off the ham? As soon as the daughter knew why the ends of the ham were cut off, she quit wasting ham.

Sometimes, there are reasons we do things. When we hook up a large diameter pony line from the engine to the hydrant, we put one full twist in the line. To the uninitiated, we do it that way because it is magic. To those with some experience, we put the twist in the line because the twist helps prevent kinks in the supply hose.

Asking why we do things helps us not to do things that don't make sense, and to do more efficiently the things that make sense.

Ego in check

The next skill the good follower has is recognizing the effect one's behavior has on others. There was an old story about a dog, a leopard, and a monkey.

A wealthy man decided to go on a safari in Africa. He took his faithful pet dog along for company. One day the dog started chasing butterflies and before long he discovered that he was lost.

So, wandering about, the dog noticed a leopard heading rapidly in his direction with the obvious intention of having lunch. The dog thought, "Boy, I'm in deep doo-doo now." Then, he noticed some bones on the ground close by, and immediately settled down to chew on the bones with his back to the approaching cat.

Just as the leopard was about to leap, the dog exclaimed loudly, "Man, that was one delicious leopard. I wonder if there are any more around here?" Having heard this and as a look of terror came over him, the leopard halted his attack in mid-stride and slinked away into the trees.

"Whew," said the leopard. "That was close. That dog nearly had me."

Meanwhile, a monkey who had been watching the whole scene from a nearby tree figured he could put this knowledge to good use and trade it for protection from the leopard. So, off he went. But the dog saw him heading after the leopard with great speed and figured that something must be up.

The monkey soon caught up with the leopard, spilled the beans, and struck a deal for himself with the leopard. The leopard was furious at being made a fool of and said, "Here monkey, hop on my back and see what's going to happen to that conniving canine."

Now, the dog saw the leopard coming with the monkey on his back and thought, "What am I going to do now?"

Knowing he would never outrun the leopard, the dog sat down with his back to his attackers, pretending he hadn't seen them yet. Just when they got close enough to hear, the dog said, "Where's that monkey? I just can never trust him. I sent him off half an hour ago to bring me another leopard, and he's still not back!"

The monkey, by working at cross-purposes to the dog, eventually got what was coming to him. Those who work at cross-purposes to an organization need to realize that saying and doing a negative thing about another person only diminishes the person saying the negative thing.

In the context of followership if a person constantly whines, like the boy who cried wolf, after a while no one listens. If a person only says something when it is important, people listen. When someone who is technically competent and well respected speaks, everyone listens. The person who is constantly negative is ignored and avoided. The upbeat, positive person is sought.

The egotistical person presents a special challenge. Whether leader or follower, the egotistical person gains a distorted perspective on the world. Since no one wants to challenge the ego, and suffer the consequences of pouting, the egotistical person only hears positive comments. This lack of action feeds the ego, which in turn draws more positive comments. The reality of the situation is usually something very different, and in the emergency services context, is something very dangerous. If no one will challenge the leader's decisions, no matter how dangerous, sooner or later the sky will fall.

A panicked person is another example of toxic contagious behavior. A tense and frightened person, noted by tone and voice, increases the tension of every other person on the crew. The more tense and out of control the frightened person is, the more the situation can deteriorate. We have all heard the sound of panic in the first arriving unit on scene. We have also heard the soothing calm of the professional command officer. The follower needs to be aware when team members are losing control of their emotions, and needs to develop skills to help the other team members regain a grasp on reality.

Establishes an authority/assertiveness balance

If there was a controversial topic that CRM creates, it is the authority/advocacy balance. Followers need to know when it is appropriate to advocate and when it is appropriate to keep their mouths shut. Several factors determine the course of the advocacy action. First, what are the risks involved? If the risks are small, the advocacy can wait until debriefing. If the risks are high, the advocacy becomes more important.

Next, the time pressures should be judged. If action is required immediately (and they rarely are on the fireground—we usually have a minute or so to think about things), like the decision whether to shock with a defibrillator, then there is little time for advocacy. However, if the patient is on a metal backboard and other people are touching the backboard, even though the time is critical, stopping others from being shocked is important.

The next item that should be weighed is the level of situational awareness. If the follower knows something the leader should know, the assertiveness should increase with the level of the critical importance of the information. If a roof is about to collapse and command has ordered an operation that would place firefighters in harm's way with no potential gain (like trying to save a couch), then the advocate should be more assertive.

If there is such a thing as good common sense, it should be the guide for when and how to use assertive behavior on the fireground. Leaders should respect the amount of courage it takes for a follower to display the assertive behavior, and the follower should be assertive only when it is important.

Accepts direction and information as needed

The good follower doesn't know everything. If the follower knew everything, the follower would be the leader. As a result, the follower should happily accept direction and new information from the leader. Any additional direction or information is a gift from the leader to the follower. The follower should always take advantage of the micro-training opportunities. Each time the follower gets more information and more skills, the follower becomes more versatile and valuable to the team.

Demands clear assignments

A good follower demands clear assignments. At a fire how often have you heard a busy command officer say to a division commander, "Go over to Division C and handle it." For experienced teams who have worked together, such a command is probably all that is needed. The team has worked together in many cases for years, and the sender of the message and the receiver of the message have a meeting of the minds on what tasks are to be performed, in what time frame, and by what people.

Unfortunately we only have that type of relationship with a few people in our lifetimes, if we are lucky. More often than not, "Go to Division C and handle it," is a completely vague command. After being given instructions, how often have you put down the radio and said to yourself, "I wonder just what in the heck command wants?"

The good follower doesn't say to himself, "I wonder just what in the heck command wants." A good follower says, "Command, I don't understand what you want. Please clarify your order."

Then the good follower mirrors the command back to the sender to insure full understanding. For example, "Division C copies. Check for extension of the fire into the Division C attic area."

Admits to errors or inadvertent omissions of assigned responsibilities

The good follower admits to errors or omissions from assigned tasks. Covering up the error or omission may in the short term avoid misery, but failure to admit the error may lay the groundwork for an error chain that has dire consequences. Earlier in the leadership chapter, we discussed how the good leader tells his followers what he does not know. If the follower knows the missing information, the follower may be able to supply that information. In the same vein, if the command officer knows what tasks have not been completed, or have been completed in error, the command officer knows not to base his strategy and tactics upon the successful completion of the tasks.

A command officer who believes that ventilation in a structure fire has been completed, when in actuality it has not, may send an entry team into a backdraft, unless the follower admits to the failure in completing the assigned responsibilities.

Additionally, reality is a scary thing. No matter how much we try to justify our actions, what happened, *happened.* When someone fails to admit to the error or omission, there are others who know what really happened. Those who know what really happened also learn that the person who will not admit to the mistake is not to be trusted. Consequently by distorting the truth, the team dynamic is destroyed, and the credibility of the firefighter who did not admit the mistake is forever damaged.

In this culture of litigation, there is a tendency to minimize and distort what happened. Our experience has shown that those who minimize, lie, and get caught, get spanked by the courts far worse than those who say, "Look, I was doing the best job I could, and in that stressful situation, the job didn't get done. I wish it had, and I did everything I could to get it done, but the fact of the matter is that it just did not get done."

Juries award verdicts against those who get caught lying. They don't penalize honest friends.

CONCLUSION

Being a good follower is more difficult than being a good leader. A good follower has all of the same responsibilities as the leader, with the additional responsibility of getting into the head of the leader, and making sure the leader's commands are followed in the way the leader envisions them to be followed. The follower has as much responsibility for scene safety as the leader has, and the authority to make the scene safe. The follower has the responsibility of knowing what information is necessary for the command officer to make informed decisions and making sure that information is supplied. The follower is responsible for insuring that commands are understood, and for helping the leader to lead. Being a follower is a difficult job, and one that carries little public reward.

Followership skills must be learned and practiced. Concentration on these skills makes the follower a more effective team member. Practice in good followership skills will make fire operations more efficient and fire scenes safer.

One final note

In some departments a culture of dishonesty, blaming, and lack of responsibility exists. Junior firefighters will not say anything negative to the leaders who have control over their careers. Leaders need to realize that negative comments on performance are a gift that allows them to improve. The follower needs to know that the comments are a gift, and to treat them as such. However in those departments where dishonesty is the prevailing culture, CRM is doomed to failure, and the department is doomed to unsafe blaming scenes where firefighters are so busy looking out for themselves that the operations and safety suffer.

REFERENCES

[1] *Wildland Firefighter Safety Awareness Study: Phase III*, "Implementing Cultural Changes for Safety," Tri Data Corporation, 1999: 5–62.

[2] Jentsch, Bowers and Salas, "Developing a Metacognitive Training Program for Followership Skills," *Tenth International Symposium on Aviation Psychology*, 1999: 338.

[3] Dunlap, J., and S. Mangold, *Leadership/Followership*, Office of the Scientific and Technical Advisor for Human Factors to the Federal Aviation Administration (AAR-100): Washington, D.C., 1998: 21.

[4] Jentsch, supra, citing Jentsch, F., C. Bowers, L. Martin, J. Barnett, J., and C. Prince, "Identifying the knowledge, skills and abilities needed by the junior first officer." *Proceedings of the Ninth International Symposium on Aviation Psychology*, Columbus: Ohio State University, 1997: 1304–9.

[5] Jentsch, supra.

[6] Jentsch, supra, citing Janis, I. L. and L. Mann, *Decision Making—A psychological analysis of conflict, choice and commitment.* New York: The Free Press, 1977.

[7] NIOSH statistics

[8] Peters, Kloeppel, Alicandri, Fox, Thomas, Thorne, Sing, Balwinski, *Human Factors*, Federal Highway Administration Turner-Fairbank Highway Research Center, "Effects of Partial and Total Sleep Deprivation on Driving Performance," Publication No. FHWA-RD-94-046.

[9] *Ibid.*

[10] *Thirtymile Fire Investigation Report*, United States Department of Agriculture, Forest Service, September 2001: 80, quoting Lamond and Dawson, 1998, whose research points out that the loss of even a single night's sleep (25.1 hours of wakefulness) impairs decision-making and vigilance to levels comparable to a blood alcohol content of .10.

[11] Under no circumstances should a firefighter respond to an incident under the influence of alcohol or controlled substances. Substance abuse has been identified in Phase I of the Tri-Data study as an ongoing but not rampant problem, particularly at incident bases and camps. Being drug-free and sober must be a tenet of the profession. Additionally, in the author's home

state, any detectable level of alcohol or controlled substance in the bloodstream, voids any workers' compensation coverage for any fireground sustained injuries, and subjects the injured firefighter to a complete loss of benefits. Alcohol and/or drugs and the fireground do not mix.

12. Welch, G. (July–August 1998) "Drink Up," *American Fitness*, July–August 1998. Welch does a very good job in this article detailing all of the effects of hydration. The article provides a brief and concise explanation of the theories of hydration.

13. Driessen, Jon, "The Supervisor and the Work Crew," USDA Forest Service, Missoula Technology and Development Center, Missoula, MT, 1990. From: Putnam, T., The Collapse of Decisionmaking and Organizational Structure on Storm King Mountain, from *Wildfire*, June 1995.

14. Introduction to Pilot Judgment, *http://www.cyberair.com/tower/faa/htm*. The Web site design is copyrighted by Stratcom Communications Corporation. The information is made available for public use and self-evaluation by the FAA.

15. Kern, T., *Flight Discipline*, New York: McGraw Hill Press, 1998:139 *et seq.* This is one of the best texts available on self-assessment and proper attitudes. Tony has brilliant insight into many things. The United States Forest Service has enlisted his participation as a human factors expert in analyzing federal wildland firefighter deaths. His analysis of the Thirtymile Fire, set forth in footnote 6 above is a must-read for any firefighter.

16. McElhatton, Jeanne and Charles Drew, *Hurry Up Syndrome*, ASRS Directline, 5, 19.

17. Della Rocco, P., E. Fielder, D. Schroeder and K. Nguyen, "U.S. Coast Guard Pilots: The Relationship Between the Home and Work Stress and Self Perceived Performance," in Jensen, R. S., B. Cox, J. Callister, R. Lavis, (eds.) *Proceedings of the Tenth International Symposium on Aviation Psychology*, 1999: 838, 840.

18. Reprinted with permission of the publisher. From *The Courageous Follower*, San Francisco, CA: Ira Chaleff, Berrett-Koehler Publishers, Inc., 1995. (This text, in the opinion of the author, is the finest available on the subject of followership.)

19. *Ibid.*

[20.] Jentsch, Bowers and Salas, "Developing a Metacognitive Training Program for Followership Skills," in Jensen, R., B. Cox, J. Callister, and R. Lavis, *Proceedings of the Tenth International Symposium on Aviation Psychology,* 1999: 339.

[21.] Chaleff, *Ibid.*

[22.] Gray, J., *Centennial Campaign, The Sioux War of 1876,* Norman: University of Oklahoma Press, 1988: 183.

[23.] Chaleff, *Ibid.*

[24.] Fischer, U., and J. Orasanu, "How to Challenge the Captain's Actions," in Jensen, R. S., and L. Rakovan, (eds.), *Proceedings of the Ninth International Symposium on Aviation Psychology,* 624 (see also Chaleff).

[25.] *NTSB,* 1982.

DECISION-MAKING

No BOOK CAN TELL a fire command officer what decisions to make on a fire scene. Tactics vary with location and the problems faced. In the southwestern United States roof operations are different than in the northern climates. The tile roofs are heavier, and the roofs are engineered to carry a smaller snow load in the southern climates. Conversely, roofs in the northern climates are typically made of lighter weight materials and are engineered to carry greater snow loads. Age of construction is also a factor. Older homes not using gusset plate trusses and typically constructed with heavier timbers can withstand heat longer than the modern "light-weight" construction. Consequently, a hard and fast rule about what tactics to use for varying conditions is destined to fail.

The same analysis applies to the wildland scenario. Some federal agencies have a hard, fast rule that people actively fighting fire do not ride on backs or tops of the units. The rule was developed out of safety considerations for the federal firefighters. However on the prairies of northeastern Wyoming, such a rule places the firefighters at greater risk than if they were riding on the engines. The fires often move faster than firefighters can run. The federal rule would place the prairie wildland firefighters at greater risk and make them practically ineffective in fighting the fast-moving, wind-driven brush and grass fires. The hard, fast rule, which works very well in one scenario, is a danger in another scenario and a rule that makes sense in one scenario may not make sense in another.

The firefighting tactics for a particular locality evolve as a result of local experience, knowledge levels, and conditions. The decisions that are made for a particular locality evolve as a result of the collective experience of the firefighters involved.

Tactics for some departments that might see a structure fire every three or four years would be appalling to departments who run 30,000–40,000 calls a year. For the inexperienced department, breaking out a window and squirting water into a room and contents fire might be all their knowledge, experience, and equipment offer them. Such a tactic to the more experienced department would be offensive and ridiculed.

Each fire officer and each firefighter must assess his or her capabilities, experience, and equipment before making a decision on how to act.

While we cannot change the training levels and equipment with this manual, we can offer suggestions on how to facilitate more effective fireground decision-making.

DECISION MODELS

For many years, psychologists have been studying the way people make decisions. They have created several models for describing how decisions are made. Most of the models of decision-making can be boiled down to a four-step process:

1. Define the problem
2. Generate a course of action
3. Evaluate the course of action
4. Carry out the course of action[1]

Essentially what the models purport is that, to make a good decision, one must figure out what the problem is (people trapped in burning building) and what the course of action should be after considering several alternatives (conduct a left-hand search from the least survivable room out, or searching the building with the use of a thermal imager). Then comes the evaluation of the course of action (will this plan work and what are the weaknesses?) and carrying out the course of action.

Given the luxury of time and a consideration of all the contingencies, the executive decision-making model can avoid the tragic consequences of a poor decision. However, the nature of the fire-fighting business is that we can never know in advance all of the strange and dangerous situations our customers will place themselves in prior to arrival on scene. When confronted with an "out-of-the-box" emergency, the models are of little practical significance in emergency fireground operations.

Gary Klein, in his groundbreaking work, *Sources of Power, How People Make Decisions*, studied firefighters, healthcare workers, and the military in determining how decisions are made under stress. The work is a must-read for anyone who would like a more in-depth study of decision-making under stress. What Klein found was that fire command officers arrived on scene, they assessed the problem, and visualized the way the problem could be solved based upon experience and resources available. As soon as a solution that appeared to work was visualized by the problem solver, that course of action was chosen. No models, no assessment, no specific steps. In emergency situations, the problem solver just imagines how the problem could be solved. As soon as the problem solver visualizes a process or procedure that will work under the given circumstances, the problem solver follows the workable course of action.

The problem solver draws upon training and experience to visualize the solution to the problem. Obviously, the greater the experience and training, the larger the supply of potential solutions to a problem. The key to success in decision-making is to give the decision-maker the most experience and training possible.

As human beings, we assess a problem by looking for patterns that match patterns already stored in our minds. If a situation matches a pattern, we tend to proceed in the way we did in the situation stored in our memory.

Successful decision-making ensures that the patterns stored in our memories are ones that are relevant to the situations we will be facing. A mismatched pattern usually leads to an unsuccessful course of action. Consequently the more patterns we have stored in our minds, the better the chance of a successful pattern match.

Additionally, by storing more patterns in our mind, we also develop informational filters, which give us the ability to prioritize pertinent information and ignore spurious information. A probationary firefighter is often overwhelmed by the volume of information contained in radio traffic on his first fire. The probationary firefighter has not yet learned how to sort through and "chunk" the information and to only rely on the information that is relevant to the particular assignment.

On the other hand, the experienced firefighter does not seem to hear any radio traffic that is not pertinent to the task at hand, but hears every word of radio traffic relevant to the task. The key to this skill is developing filters that sort the relevant information from the irrelevant information.

We can increase the patterns and filters stored in our minds through experience, training, briefing, and debriefing. We can pyramid our stored patterns by carefully listening to the experience of others. Our mind does not differentiate between patterns we have learned through storytelling and patterns we have learned through experience.

Detailed storytelling has the same effect as experience. If we use storytelling correctly, we can gain experience even if we are not personally at a particular incident.

THERE IS A SUBSTITUTE FOR EXPERIENCE

Nothing prepares a firefighter to face a situation better than having faced the same or similar situation successfully. The more experience the firefighter has, the better the decision-making pool of information to draw upon.

For example, the Fairfax Virginia Fire Department runs lots of automobile extrication calls. They have a large volume of car accidents, and they are one of the best departments in the world in getting patients out of the car, and on their way to the hospital in the shortest possible time. Individually, the Fairfax Rescue Technicians have great rescue knowledge. Collectively the Fairfax Fire Department has a treasure trove of automobile extrication experience.

It doesn't come as a surprise then, that the Fairfax SOPs for automobile extrication are sophisticated, efficient, and well developed. A Fairfax command officer will run an efficient automobile extrication scene. Fairfax has drawn on the collective experience of its department to establish a state-of-the-art automobile extrication program.

While there is no substitute for on-scene experience, using real-life people and situations, with real-life consequences as a training ground for decision-makers is at best a risky proposition and at worst a disaster.

There are techniques that can be used to enhance a decision-maker's ability to make the right call in a stressful situation. Consistent use of the following techniques will result in better decision-making.

TOWARD BETTER DECISION-MAKING

Unfortunately we all can't be everywhere all of the time. Before we are faced with a decision, there are some steps we can take to fill the patterns and filters in our minds with information, which will help us to be better decision-makers. We can train, debrief and tell stories. By creating "artificial experience," we place more tools in our decision-making toolbox.

Training

No firefighter on the first day on the job is turned loose with an engine and told, "Here is your engine, drive it to the fire, pump it, and put out the fire." Either through a formal training process or through an apprenticeship program, firefighters are taught how to fight fires. The novice firefighter learns about fire behavior, throwing ladders, pulling hoses, and fighting fire. Before firefighters ever put on an air pack, they have had training on what to expect from the fire, how to don and doff the gear, proper operations of the equipment, and proper safety procedures.

What can the individual firefighter do to make better decisions? Simply put, the firefighter can gain more experience in an artificial environment by training. The key to effective training is to follow the old Vince Lombardi philosophy, "Train like you play."

In stressful situations, a firefighter reverts to prior over-learned behavior. If a firefighter trains with bad habits, the bad habits become over-learned, and when placed in a stressful situation, the firefighter reverts to that behavior. In one department with which we've had the opportunity to train, the firefighters do not go on-air in rescue training situations. Instead of refilling the air bottles, the firefighters train with the air packs on, but with the air disconnected.

While the firefighters get the opportunity to practice rescue skills knowing how the equipment feels and reacts during the rescue drill, the shortcut of not going on air limits the learning experience.

The firefighter does not get the opportunity to learn about air usage and budgeting air. Unrealistic time expectations are created. Moreover the firefighters develop the bad habit of going off air while in an imaginary hot zone.

One time in a ceiling collapse situation in a real fire, one of the firefighters of this department reacted to the stress of having the drywall and insulation come down on his head by going "off air" just like he had in drill. He got a face full of smoke, a ride out on the shoulders of his partners, and a trip to the hospital in an ambulance as a reward for his bad training habit.

The lesson is to train like you play. Make sure you simulate the real world as closely as you possibly can, while at the same time insure the safety of those under your charge. Training should be a time to make mistakes and not to get hurt—not a place where error-free performance is demanded.

We've all been to the training drills where someone stands up in front of the class and says, "This is the brand new Mark VII Widget. You use it by plugging the green hose into the red fitting, pulling the lever forward, and twisting knob *B*. Make sure you don't twist knob *B* too far or the alarm will go off and the widget's lining will explode, leaving feathers and tennis shoes all over the street....All right. Any questions? You all got it? Good. Use it in the field. Let's go get some coffee."

That type of training can be absolutely useless in a stressful situation. Those skills, which need to be automatic in a stressful situation, should be rehearsed until they *are* automatic. One particularly innovative fire instructor at his fire academy requires his students to tie 100 bowline knots at the beginning of each academy class. This instructor knows that the ability to tie a bowline automatically may be a life-saving skill some day. Then the instructor places the students under pressure, by requiring them to tie the knot in 10 seconds, then 5 seconds, and finally 3 seconds. The artificial time pressures simulate the same type of reaction that results from on-scene emergency stress. The repeated training causes the knots to be automatic. This instructor knew intuitively that training like you play will save lives later on in the firefighter's career.

The U.S. Military did extensive studies of team decision-making in its Tactical Decision Making Under Stress (TADMUS) program.[2] As a result of the TADMUS research, the development team found those teams that perform the best under stressful situations are those teams that adapt their decision-making strategies, coordination strategies, and even their structure when faced with escalating workload and stress.[3]

Based upon the research contained in D. Serfaty's article, "Team Coordination Training," we learned that adaptive teams are high performers, and that adaptability is trainable.[4] The researchers found that the following elements are necessary for successful team adaptability. First, the team must all have a common mental model of the situation.[5] The command officer, routinely giving his assessment of the situation, facilitates sharing this information.

Additionally, the level of communication of relevant information increases in the adaptive teams. The team develops common language and shorthand communication techniques. An example of the high-level team interaction would be the Fairfax Fire Department Rescue Team. The rescue lieutenant has developed a series of hand signals to command his team on automobile extrication. The hand signals are easily visible and overcome the loud on-scene noise that is usually at the fire scene.

Finally teams that are successful exhibit anticipatory behavior. The team members know what the commands are going to be, before the command officer issues them, and they mentally prepare themselves for a course of action. This behavior is a product of the other three steps, and is the most effective when the other steps are effective.

To take advantage of these team efficiencies, a team should train together often in lifelike situations. The more the team trains together, the more the team advantages are developed.

Debriefing

If properly conducted, debriefing can be one of the single greatest teaching devices available to a fire command officer. If not properly conducted, a debriefing is a monumental waste of time. Properly debriefing an incident immediately after the incident while memories are fresh, allows all of the participants to get the entire picture of how an incident came together and how actions by one division interrelate with actions performed by other divisions of a fire. Events experienced by one group of firefighters can be shared with other firefighters.

While this issue is addressed in another chapter, a comparison between a poor debriefing and a good debriefing is important for a complete understanding of the process.

A well done debriefing includes a retelling of the events as they happened. Human beings learn by hearing stories. The stories provide content, relevance, and the emotional ties necessary to remember and relate to the information and learning experience. Building a time line of what happened and how it happened at the beginning of the debriefing provides relational content upon which to build the rest of the debriefing.

An example of a good debriefing

The tones went off. Command arrived on scene two minutes before the first due engine, Engine #11. Command walked around the burning structure and determined there was a room and contents fire in the kitchen. The reporting party informed command that all persons were out of the house, and had been accounted for, but the family dog was still inside the home. Command told Engine #11 it was first due rescue, and to look for the family dog before starting suppression efforts. The second due engine was Engine #31. Engine #31 arrived on scene two minutes after Engine #11. Engine #31 was assigned first due suppression, and they tagged the hydrant, and provided water supply.

Jones and Smith began a quick interior search using the thermal imager for the family dog. They found the dog in the basement, and brought it out. EMTs on scene did doggy CPR and revived the dog.

Swenson and Barlow did quick knock down. Some of the things we did right were good communication, a quick knock down, and we saved the dog and the family property.

We could have had a Rapid Intervention Team (RIT) team established sooner.

An example of a poor debriefing

We just finished the fire. We did a good job. Anyone have any comments on what we did right? Okay. Anyone have any comments on what we did wrong?

Use of the debriefing checklist provided in the debriefing chapter will facilitate debriefings.

Training can be a source of information, which increases the decision-making ability. Good training can be the source of good information upon which to make a decision.

Preplan

The next step we can take toward better decision-making is to preplan an incident. By looking at an incident and saying, "If *A* happens, then we do *B*." We take the decision-making out of the time-stressed environment. We move it to a more analytical environment in which options, including novel ones, can be applied to the facts at hand to reach a decision. The preplanning process also starts the problem solving mechanism of the brain, long before a problem is reached. For those with limited experience, the preplanning exercise gives the novice firefighter the opportunity to ask questions outside of a time-stressed environment.

Preplanning can be formal or informal. We all know about the formal type of preplan, which includes visits to buildings, inspections of the premises, listing of assets and hazards, and formalization of an attack plan before anything bad ever happens. It goes without saying that formal preplans are one of the most useful tactical tools we have in the fire service.

The formal preplan also serves another, more intangible function. By preparing a preplan, we force our personnel to think through the hazards of a particular location without the stresses of time we have on-scene. We develop the thought processes and habits necessary for effective fire command by rehearsing them repeatedly in preparation of several preplans. All people who are responsible for making tactical decisions (basically everyone who puts on a uniform) should participate in the preparation of formal preplans.

After the plans are prepared, they should be critiqued from a tactical perspective. Given the potential situation, what would work better and why? By such formal discussion and give-and-take, we train our firefighters in the tactical decision-making process long before they ever see a similar situation. By putting our people in imaginary "real-life" situations and allowing them to think their way out of the situation, we develop more able, creative, and efficient firefighters. By discussing the plans as a crew, we develop cohesion and we also learn the way other people on the team tend to make decisions, allowing for anticipation and synergy.

On the other hand, there are lots of opportunities for informal preplans. In our department we have occasion to fight lots of wildland fires. Just within our county, travel times to a fire may be as long as two hours. We have made it a habit on the way to the fire, to discuss the type of fuel in which the fire is burning (in our county it can range from swamp grasses, to desert, and timber), the fuel moisture, weather predictions, and recent fire behavior. If a front is predicted to pass sometime during the day, we might discuss when the frontal passage is expected, and what we plan to do if we see the front coming across the fire.

While our preplans are theoretical, they cause us to think about the situation and to get our minds on the problems. We find it to be more productive time than spending the two hours on the way to the fire discussing local politics or the sports page.

The same analysis applies to the structure fire. In the three to four minutes it takes to get on-scene, each individual or the crew can run a mental preplan. For example, en route to a strip mall fire an individual can think to himself things like, what type of occupancy do we have? What are the rescue considerations? What type of construction do we have? Does the construction pose any special problems? Have we fought a similar type fire, and if so, what were the problems we had? What type of help are we going to have, and when is it going to arrive?

Novice firefighters should be included in preplanning activities because they are given the opportunity to assess situations and tactics long before they are confronted with the overwhelming task of making decisions at the fire scene. The preplanning process is a micro-training opportunity, which allows the senior firefighter and the novice to test each other and to exchange knowledge of tactics and operations in a safe situation.

Storytelling

One of the most effective training programs that exists in every fire department is the storyteller. Human beings all learn by the telling of great stories. All of the great nations of the world have taught their young by the telling of stories. The Sioux Indians taught bravery by telling stories of great deeds. The great religions of the world teach morality by the telling of stories. We all teach our young to be careful, safe, and frugal by the telling of fairy tales. Also, we teach our firefighters to become better firefighters by the telling of stories.

Those boring old firehouse stories that get told over and over again are a way of sharing our experiences with other firefighters. The subconscious does not differentiate between reality and fiction. Through the accurate use of storytelling, we can increase our ability to respond to fire scenes.

The following is a story that illustrates the effectiveness of storytelling.

The dispatcher told us we had a working structure fire out in the boondocks of Campbell County, Wyoming. The house was 12 miles from the nearest response engine. As was typical for this rural area, there was no water supply for miles. I was on the second due engine, and was assigned to set up a supply engine for the tanker shuttle. The first due engine dropped a large diameter line at the beginning of the driveway and laid the line all the way down the driveway. There was no way to get tankers in and out of the driveway in any efficient way, so the use of a supply engine was mandatory.

I set up the drop tanks from the tankers. The tankers dropped their water into the tanks and went off the five miles to the nearest water supply. My crew assembled the hard suction line and put it into the drop tank. We pulled the primer pump. No results. We tightened down all the connections, and tried to prime the pump. No results. The first due engine was screaming for water. If we didn't get water, we were going to lose this house.

We looked at the hard suction. It was obvious there was an air leak in the system somewhere. All the bleeders were closed and the connections tight. We pulled the primer and listened. Then, we heard it. There was a pinhole in the hard suction, which was allowing the pump to suck air.

The next closest piece of hard suction was in town, at least 20 minutes away. If we waited for the hard suction, we were going to lose the house.

Then I remembered a story an old oilfield hand told me. He called his technique, "oilfield technology." When they had a leaking hose, they took an old piece of inner tube and wrapped the hose. It was not the best practice, but the old oil field hand used to say, "Out in the oil patch you can't call a plumber to fix every leak in the toilet."

We didn't carry inner tube sections on our fire engine. We inspected our equipment routinely, and we expected it to work the first time—every time. Sometimes though, you have to adapt and overcome.

I asked myself, "What do we have on a fire engine that will work like a section of inner tube? We carry all kinds of suppression gear. We carry all kinds of rescue tools. We carry medical kits."

I stopped. There was something about medical kits. Then, it came to me. I pulled out two latex rubber gloves. I tied them together, and then I tied them around the hard suction line, over the leak. Then, I took tape and taped the gloves in place."

We pulled the primer. The pump primed. We sent water up the supply line and we saved the house—all because an old oilfield hand bragged to me about "oilfield technology."

The story teaches us to adapt and overcome. More than that, it places a pattern in our minds. The pattern is that if there is a leaking suction line, use something to plug the leak. It is not a recommended practice—high quality hard suction line is the recommended practice—but given the stakes and the situation, adaptation saved a home.

Even though only a handful of people were on-scene and even fewer were involved in the solution, this knowledge and experience will become a pattern in the minds of many more people.

DECISION-MAKING AIDS

As we have explored earlier, the human machine is limited in the capacity of information it can process in a given period of time. As stress increases, the ability to receive and process information decreases. These limits of decision-making ability can be extended through the use of decision-making aids. One thing we know for certain, the decision-maker is not going to get any smarter between the station and the incident, and in most cases, the decision-maker is going to get dumber. As a result, the decision-maker should develop a cache of decision-making aids to facilitate fireground decision-making.

The checklist

Anytime a passenger hops into a commercial airline to take a trip across the country, that passenger can sit back and relax knowing that the plane will not crash because the pilot simply forgot to flip an important switch. The flipping of the switch is not missed because of the checklist. Even though a pilot may have 12,000 hours flying airplanes and may have made 25,000 take-offs and landings, that pilot is still required on every flight to use a checklist. All of the takeoff procedures are listed, step-by-step on the checklist.

The consequences of not flipping the right switch can be disastrous. Wherever that airplane is going, the pilot is going to get there first, whether it be to the destination, or into the ground. So the use of the checklist as an error prevention device is widespread, common practice in the aviation culture.

The fire service can benefit from the use of such checklists. An example of the use of a checklist that is in common use in the emergency medical service is the SAMPLE history.

S – Signs and Symptoms

A – Allergies

M – Medications

P – Past Pertinent Medical History

L – Last Oral Intake

E – Events leading up to the onset of the problem

The mnemonic device SAMPLE helps the first responder to remember the checklist of important information that the treating physician needs gathered to provide effective healthcare.

Another example of a checklist, in widespread use by the wildland fire community is LCES.

L – Lookout

C – Communication

E – Escape Routes

S – Safety Zones

No wildland firefighter should commence any firefighting action before establishing all of the elements of the LCES mnemonic checklist.

Yet another example of a checklist in widespread use by the Hazardous Materials (HAZMAT) community is SIN:

S – Safety First

I – Isolate and Deny Entry

N – Notify the proper authorities

SIN is the mnemonic device that is taught to first responders when encountering a hazardous materials incident. It is an easy word to remember, and quickly and easily sets forth the steps to be taken at an emergency incident.

The structural fire service could benefit from the use of more checklists. In much the same way as the first responding emergency medical technician gathers information for the treating physician with a SAMPLE history, the interior structural firefighter can be the eyes and ears, and relay the necessary information to command. The following mnemonic checklist is a guideline of the information that should be relayed to command.

S – Structural Stability

A – Access In

F – Fire Stage/Behavior

E – Exit Out

S – Smoke

T – Temperature

O – Occupancy

P – Potential Fuel/Fire Load

The SAFE STOP mnemonic device can aid the interior firefighter on what information command needs to make strategic decisions. The purpose of this manual is not to provide tactical information for fighting fires, but to describe the human machine's interaction with the fireground and the firefighting equipment. This mnemonic device will make the firefighter's interaction with command more relevant and informative. The consistent use of checklists on the fireground will make firefighting decision-making much easier.

Timers

Another decision-making aid that is useful on the fireground is the timer. Tactical decisions are time dependent. Unfortunately in the heat of an emergency situation time perception gets distorted. Sometimes the perception of time stretches out. Sometimes the perception of time shortens. The reality of situation is that time continues, second after second, passing at the same rate. However our perception of time varies with the situation. Whether we are on a fireground or in a spaceship traveling near the speed of light, our perception varies.

To counteract the dangerous effects of time passage, many departments are purchasing large timers that are attached to the side of the command unit. As soon as the call is received, the timer begins. The large timer is positioned so everyone on the fireground can see how much time has passed into the incident.

Structural degradation, crew endurance, and fuel supplies are all time-dependent. Tactical decisions are time-critical. The timer is a device that insures that the command officer and the individual firefighter have vital information necessary for operation on a scene.

Some departments use the dispatch center as the timer. The dispatcher calls out to command every 10 minutes from the initiation of the call. The advantage of the dispatch timer is that everyone listening to the appropriate frequency knows how long the fire has been burning.

Officer's aide

The command officer's aide has been used since the advent of organized firefighting. Recently there has been a renewed interest in the officer's aide. We know from earlier chapters, that the human mental capacity is limited. The officer's aide is simply an extension of the command officer's memory.

On another level though, the purpose of the CRM program and of this text is that every person on-scene employed properly, is an officer's aide. All of the skills set forth in this text are designed to develop every person on the scene as an aid to the successful completion of the mission.

THE DECISION-MAKING ENVIRONMENT

Finally, something must be said about the decision-making environment. Good decisions cannot be made in an atmosphere of chaos. Disciplined communications with relevant information are the key to a good decision-making environment.

The aviation industry has developed a protocol for critical times in flight operations. It is called the "sterile cockpit principle." The rule in the aviation industry is that during critical times in flight, the only matters that can be discussed are the tasks at hand. All irrelevant communication is prohibited. No one talks about their spouses or significant others or the good time they had last night. Instead all communications—and in theory, all minds—are focused on the task at hand.

The fire service can learn from the experience of the aviation industry in this regard. During critical times in the operations, communications, both on the radio and off, should be focused on the task at hand. By focusing the communications to the task at hand, the fire service can minimize distractions. Firefighters are focused on tasks to be performed and situational awareness. The environment for making good decisions is created.

We offer the following challenge to our readers. Without telling the engine company crew why they are being taped, place a video camera in the cab of an engine and videotape a few responses. Then watch the videotape and determine how much of the discussion during the critical time on the way to the call is non-operational discussion. Then, limit the discussion to operational-only issues and see if the operation goes more smoothly during those critical first few minutes of the operation.

KEYS TO GOOD DECISION-MAKING

In addition to the techniques discussed earlier in this chapter, we believe the keys to good decision-making can be boiled down to the following four simple steps.

1. **Maintain good situational awareness.** Make sure you know what the problem is, the factors influencing the problem, and the anticipated consequences of each course of action. Most importantly, assess the time available to make a decision. Will a small correction now, save a lot of work later? Is drastic action required now? Is there time to think about things before we move?

    One of the major temptations on the fire scene is the, "Do something now!" demand. After all we are the emergency services. In an emergency we do something, even if it is busy work; it is something. Often the best course of action is to assess the situation and sit back and wait.

2. **Maintain technical proficiency.** You are not any good as a decision-maker if you don't know how things happen and why, both from a technical perspective and a CRM perspective.

3. **Know your resources.** Knowing what tasks your personnel and equipment are capable of performing, avoid giving impossible assignments that damage morale, credibility, and risk the operation. Know pump capacities, water supply, personnel training, scene hazards and all the other myriad things that the fire service expects of its decision-makers. If you don't know the information, make sure you ask someone who does.

Recently a tank battery was hit by lightning. One oil tank exploded and another was flame impinged. The command officer told the personnel to pull a $2\frac{1}{2}$" line and put foam on the fire.

The engineer on the scene met privately with the command officer. He asked, "Don't you think it would be better to use the deck gun on this fire?" The command officer said, "No, I want a $2\frac{1}{2}$" with foam." The engineer said, "I would be glad to do that, but this engine does not have foam capability on the $2\frac{1}{2}$". Don't you think it would be better if we used the deck gun that has foam?" The command officer told his crew, "Let's use the deck gun with foam on this fire."

The engineer advocated his position privately and without embarrassing the command officer. The engineer knew the capability of his engine and was the subject matter expert. The command officer relied on the engineer as the subject matter expert and adjusted the tactics accordingly.

4. **Evaluate, evaluate, evaluate.** Just because you have made a decision to proceed one way, does not mean you are locked into that course of action if the tactic is not working. Evaluate your course of action. If it is not working, do something else. The definition of insanity is doing the same thing over and over again and expecting different results.

On-Scene Tips

We offer the following tips and techniques to facilitate better decision-making. First, pre-load your people with patterns and filters. For the same reason you would not go into a war with an unloaded weapon, you would not go into a fire with an unprepared crew. Talk about past fires. Do table top exercises. Train, train, train. Study the successful actions of others. Give yourself the best weapon you can obtain to fight the emergency situation.

Second, use notes, operational checklists, status boards or other tracking devices or tools to reduce the mental load you are carrying. Know what you need to know and know where to find the rest. Keep your working memory free to consider relevant facts. Don't clutter your mind with minutia.

Third, learn to recognize when you are overloaded mentally and physically. Teach others to recognize those conditions too. There is no shame in asking for help when the alternative is risking a mistake that will hurt or kill someone. Make it okay for other members of your firefighting team to tell command or other firefighters that they are losing the edge and that they need to take a step back. Make it okay for anyone to say, "This just doesn't feel right. Can we think about what we are doing?" Respect those gut feelings. They come from subconscious pattern matches. The body tells us things for a reason that sometimes our mind does not yet know. Listen.

Finally, don't resist help. Help is good. Help makes things easier. *Handling it, squeaking by,* and *disaster* are all close neighbors. Don't invite them to your scene. If nothing else, you can use the helping opportunity as a micro-training opportunity to teach someone else the skills of command and decision-making.

Conclusion

Pattern recognition is the key to effective decision-making on dynamic emergency scenes. Through increased training, debriefing, pre-briefing, and storytelling, the decision-maker gains additional patterns upon which to draw when faced with a decision. There are additional memory devices, which extend the natural human abilities. Employing those memory extenders will also help the decision- maker to manage necessary information. Effective decision-making is a skill learned over time.

Wrong decisions are made when there is a pattern mismatch. All of the pattern creation exercises should be crafted to create a situation as real life as safely possible, so the chances of pattern mismatches are avoided. Poor training carries over into poor decision-making.

Finally, decision-making is a learned skill. Those potentially in decision-making positions should be trained to make a decision. Inexperienced decision-makers make inexperienced decisions. Every firefighter should be trained to a level of proficiency, and then placed in safe situations in which decisions need to be made. In that way decision-making skills are increased.

References

1. Klein, G., *The Sources of Power, How People Make Decisions*, Cambridge: MIT Press, 1999: 127. Also citing, Lipshitz, R., and O. B. Shaul, (1997). Schemata and mental models in recognition-primed decision making. In Zaambok, C. and G. Klein (eds.), *Naturalistic decision making* Mahwah, NJ: Erlbaum, 1997: 293–304.

2. Collyer, S., and G. Malecki, *Tactical Decision Making Under Stress: History and Overview*, in Cannon-Bowers, J. and E. Salas, (eds.) *Making Decisions Under Stress*, APA: 1998.

3. Serfaty, D., Entin, E, and J. Johnston, *Team Coordination Training*, in Cannon-Bowers, J. and E. Salas, (eds.) *Making Decisions Under Stress*, APA, 1998: 222.

4. *Ibid.*, 242.

5. Orsanu, J., *Shared mental models and crew decision making*, Paper presented at the 12th annual conference on the Cognitive Science Society, Cambridge, MA, (1990) as cited in Serfaty, D., E. Entin, and J. Johnston, *Team Coordination Training*, in Cannon-Bowers, J. and E. Salas, (eds) *Making Decisions Under Stress*, APA, 1998: 240.

DEBRIEFING AND CRITIQUES

WE SPEND HUNDREDS and hundreds of hours every year training and retraining firefighters. We set requirements for initial and recurrent training. We schedule classes hourly, daily, weekly, monthly, yearly, biennially, and on and on and on. We obliterate weekends and other days off. We pay overtime for students and instructors. Why?

1. To replenish the knowledge that we supposedly forget every year.

2. To replace the skills that we somehow lose, even though we continue to perform adequately incident after incident.

3. To take the place of missed opportunities that we don't use after every incident.

What we should be doing as a fire service, is to use scheduled training for teaching new skills, and incident debriefs to refresh our knowledge. Also, we must employ refresher training, not just annually, but whenever sub-standard skills exist as a whole.

Currently most refresher training is not tied to any skill level, ability, or currency. It is tied to a subjective time period after which time firefighters "must" have lost some of the important knowledge and need re-introduced to the same class over and over again. The current philosophy is that if we do not require firefighters to attend the refreshers, then they will have a problem performing safely. Nothing wastes firefighters' valuable time more than making them sit through training that does not apply to the way they operate or retrain them in something they are already proficient at and use often in their daily jobs.

What the fire service should be doing is lobbying *against* mandatory refresher training based upon subjective time frames. What the fire service should be doing is canceling refresher training. What the fire service should be doing is devoting their precious dollars and hours to training that will make a difference in the safety of their firefighters and the efficiency of their operations. Mandatory refresher training wastes the instructor's time, the firefighter's time and the citizen's dollars. Saving "a problem" for refresher training is a disservice to our firefighters and our public we protect.

A more valuable type of refresher training should be taking place after every call. In this way, training can be fresh, pertinent, meaningful, timely, applicable, and useful. Some departments already do a great job of debriefing or critiquing an incident immediately afterwards, and it is like money in the bank for those departments. But some departments do not debrief their incidents, preferring to ignore the issues or to save them for a later date.

This chapter will help your training program by making it more flexible to the needs of the department. It will help your firefighters by teaching real world examples. It is an opportunity to increase your team's CRM proficiency and re-emphasize those principles. Also it will make your department utilize its training dollars more efficiently, by identifying where they need to be spent, and not just where some bureaucrat says money should be spent.

THE MISSED OPPORTUNITIES

We should review our incidents and actions every time. These are the best opportunities for transferring important information to other firefighters, identifying strengths and weaknesses, and building perfect practice for perfect incidents. It's sort of like refueling our mind. By ignoring these opportunities we are wasting a valuable resource—the fuel of knowledge. You respond to a call. You perform your jobs on the incident scene. Then you return to the firehouse and drop in front of the TV or go home. We have spent all this time and effort for what? We have just wasted a very precious resource and squandered an opportunity to use this commodity for a good purpose.

Reviews should be performed after every incident—large or small. There is something to be learned from every single one by every single firefighter. Maybe you will not learn a new skill, but maybe you will learn something about your partner or your officer or human nature in general. But if you really look hard and try, you will learn something from every incident. Better still you may teach someone something that will help them in their jobs or maybe even prevent their injury or death.

DEBRIEFS, CRITIQUES, AFTER-ACTION REVIEWS, AND OTHER WHAT-CHA-MA-CALL-ITS

Yes, it does matter what you call it. Not because we do not want to hurt anyone's feelings, but because it sets an expectation of the discussions. Critiques tend to have a negative connotation, and are many times instructor led. Debriefs are generally more group oriented. Should you choose one over the other for your department? *No!* Use them both. Critiques have their place in a department and a team environment. Critiques are more of an instructional opportunity with reality as the lesson plan. An instructor, officer, or other leader gets up in front of everyone and goes over what happened and what needs to be fixed. There is not much group interaction in this type of review.

Things at this point have gone so horribly wrong that a review of the incident by the group is not warranted. It takes too long to discuss what all went wrong in the group discussion mode. By judging from the performance of the team, one might conclude that they may not have the expertise to conduct an adequate critique anyway. This method borders on a lecture. People made major mistakes that need to be fixed—no bones about it. No need for discussions, just training—or retraining.

These critiques can occur for various reasons. Maybe the crews that handled the incident have little or no experience or training in this type of call and were not able to adequately perform. Such is the case with unusual rescue cases, hazardous materials, unusual fires, etc. It would be pointless as a facilitator of a critique, to attempt to guide a discussion of what to do when Methoxy-butyl-3-6-9-acetylgiveacare is on fire, when no one knows what the heck it is! Therefore, an instructor-guided tour of pertinent SOPs, strategies, tactics, resources, etc., to be used in this type of incident is probably best. A critique is most useful in these types of incidents.

On the other hand, debriefs are also valuable tools but are not based on one instructor teaching the group. Debriefs are discussions of the incident and what we can learn from them. It is guided by a facilitator, not lectured by an instructor. These debriefs take place when people are knowledgeable about the type of operations that they were engaged in, and therefore have significant information to contribute and transfer to the team.

THE ATMOSPHERE IS
"THE BLUE LINE"

You would not think that jet pilots and firefighters have much in common. But we could learn a lot from a group of jet jockeys called, "The Blue Angels." These highly trained and disciplined professionals place their lives on the line every time they are in the air (wait—that's like us). Their aerial routines are so precise that as much as a couple of feet deviation could spell disaster for themselves and their fellow pilots (us, too). Therefore they have designed and practiced a very progressive method of insuring everyone is performing at near perfect levels (we are not so formal, but we are working towards this).

Of course being military, this group has a very specific rank structure (us). Military officers of this caliber are to be treated with the utmost respect and courtesy (yep, us). After every performance, the extraordinary pilots of these jets enter a debriefing session that is totally contrary to normal military, or any other rank-based organization. When these military officers enter the debriefing room, they are stripped of all rank. They enter this room as individuals whose lives depend upon one another, not as a someone with more authority or rank than the other.

If the squadron commander made a mistake during a performance, it is actually thought to be bad form not to bring it up for discussion. All aspects of the performance are reviewed. All aspects of individual performance are discussed. When the Blue Angels exit the debriefing room, they exit as better pilots and a stronger team than when they entered. This is where the concept of the "Blue Line" enters into the fire service.

We should learn from the Blue Angels' example of openness, honesty, self-improvement, and team building. To set the stage for attaining this level of openness, every department should take a roll of blue tape or blue paint and make a line across the entry door to the training room (or wherever debriefs and trainings will be held). After this, place a banner across the doorway that says, "When you cross the Blue Line, we are all equal." Inside, and during the introduction to a debriefing session, make the rules of the Blue Line known to everyone,

Rules of the Blue Line

- This is an open, non-aggressive atmosphere to discuss the relevant issues at hand.

- No prejudices, defensiveness, or hurt feelings will be allowed either during the debriefing or after we leave.

- You are expected to openly discuss issues that are relevant to the debriefing.

- What you learn here will make you a better firefighter and may save your life someday.

- What you say here stays here.

When people leave, another banner above the exit should say, "Learn from what is said behind the Blue Line."

Basically, if we cannot be expected to speak up during a debriefing, we cannot learn, others cannot learn, and we all cannot improve and be safer. If discussion from everyone is not expected, you will only hear from the vocal minority—some of which will not always be correct in their vocalizations. Make it known that the debriefing atmosphere is without rank, without anger, without defensiveness, but professional and respectful. Also, make sure that speaking up is an expectation and not the exception. This way we can learn as much as possible from every incident we handle.

THE PATHWAY TO ENLIGHTENMENT

There is a certain way that works best for conducting debriefings. Of course there are many different opinions on the format of a debriefing, but we have found that this particular format works best in our world of firefighting and human behavior associated with risk-takers.

General debriefing concepts

To conduct a proper debriefing of an incident it must be

1. Within a few hours after the incident

   a. **Exception.** Excessively fatigued personnel cannot be debriefed effectively. In fact, there should be a couple of strong bouncers if you try this. Physical and mental fatigue will predestine firefighters for problems in accepting feedback, properly giving assessments on the performance of others, and will result in an overall ineffective debrief and maybe a scuttled team.

   b. **Exception.** Extraordinary events preclude an immediate operational debriefing. Serious injuries, firefighter fatalities, civilian fatalities, and large operational scope, etc. may mean that other actions must take place prior to debriefing the actual operational effectiveness of the incident. Critical Incident Stress Debriefings (CISD) are not covered in this chapter, but are of utmost importance in the mental and physical health of your firefighters. When events occur on this level, CISD, information gathering, planning/scheduling, investigations, etc., are items that may need to be addressed prior to the actual debrief.

   Do *not* skip the operational debrief because of a significant emotional event like a line-of-duty death. After your firefighters' emotional health is addressed, the operational debrief may help to lend a sense of "making a difference" to those involved. If we could talk to those firefighters killed in the line of duty, I truly believe, they would say that they do not want their deaths to be in vain or the same problems to kill others. "Learn from my death so I may teach others." I think they would want that.

   To dispense with the operational debrief of a tragic incident is a tragedy multiplied. Questions go unanswered by those who could answer them and are left to the destructive armchair quarterbacking at the coffee pot. Personnel who need the information to keep them from repeating the same errors are ordained to the same fate. Then maybe those individuals, who require the most attention operationally, are not identified. "Those who cannot remember the past are condemned to repeat it," said George Santayana. A department has to have in its culture this desire to learn from their mistakes, to build on their successes, and to create a stronger team.

2. With everyone involved

    a. **Exception.** If personnel are not able to attend the debriefing, an officer must make the effort to meet with the absentees to address their concerns and issues and to transfer the knowledge gained from the official debrief.

3. Of a sufficient length to include available feedback and discussion

    a. **Exception.** Time constraints prevent this occasionally. Make an effort to meet the objectives of the debriefing while including as much discussion and feedback as possible.

Debriefing template

The actual debriefing is conducted in the same general format for every incident. After a few, the firefighters are aware of how debriefings function and are better prepared to conduct one.

Step 1—Just the facts. First, we discuss the actual incident facts. This is an account of the initial call-out, the resource assignments, the actual incident, individual assignments, etc. This is the starting point for all discussions. It is helpful to remember that not everyone heard the initial call-out to a car fire, and the subsequent response of one engine company. They only know that the idiots in town sent one engine to a fully engulfed garage fire attached to a dwelling. By reviewing the facts of the incident, leaders can answer many questions, and temper, er— many tempers.

Step 2—What did you do? Arguably a portion of the "Just the Facts" section, the "What did you do?" portion, gently opens the door for everyone to explain what their assignments were, what they did, what they encountered, etc. Although difficult, and darn near impossible, we attempt to steer clear of the good, the bad, and the ugly in this discussion. This again is fact-based and does not cover the "I think" items.

Step 3—What went wrong? This section is what all firefighters are here for. They are here to grab a chunk of someone's backside for whatever reason. While some debriefings attempt to steer clear of this section or to sugarcoat it, we specifically address the issues. We set the stage in the first two sections by clearing up any misunderstandings that may have been present on the incident. Not everyone knows what Engine 31 was doing at the rear of the building. Not everyone knows why we turned the electricity back on. By addressing the facts initially, a leader can explain what seemed to be an unjustified action on-scene that becomes an understandable order when the facts are known.

We allow firefighters to address those issues, tactics, etc., that they do not understand or do not agree with. By emphasizing non-aggressive, honest discussions, and the Blue-Line Concept, we attempt to bring out those mistakes and misjudgments that could eventually kill someone. We do not sugarcoat a mistake in the debriefing. If you messed up you messed up—expect it to be noticed by others and discussed. No harm, no foul; just do not do it again.

In the old method of critiques and debriefings, we often addressed the negative results of an individual's actions and mistakes. But we hardly ever heard, "Then Harland went to the back door and kicked it in, increasing the fire and causing additional damage." What we often heard was, "*Someone* went to the back door...." In a small department or operation (or even if one other person knows), that someone is quickly identified and "strung-up" by the coffee pot.

This is unfortunate because someone does not have the opportunity to address the actions and reasoning. It's sort of the American philosophy of "Innocent until proven guilty" thing here.

Bottom line is we cannot get better as individuals, as a team, or as a department if we do not address sub-standard performance openly and honestly. Remember the concepts and rules of the Blue Line.

Step 4—What went right? Yes of course, we do things very well on most incidents; we just never really talk about them. Isn't that a shame? We perform heroic acts practically every day. We risk our lives and our health. For what? We should at least be recognized for what we do right. Goodness knows that we get recognized for everything we do wrong!

We save this part for the end because it helps to form a closure to the incident and the debriefing, and re-unites the team. After we finish going over what went wrong and everyone gets bent out of shape, we end on a positive note. We give people a chance to review everything that we did right. Did I just say opportunity? We *make* everyone recognize the good things that happened. Our scene may have been tactically the worst anyone has seen in 30 years of fighting fire. But strategically the fire went out, no one got hurt, and we all went home to our families.

Many arguments during operational debriefings are based on the *tactics versus strategy* of the incident. Example. On a medical call maybe we could not get a splint on, the tools would not operate, backup never came, the portable oxygen tank was empty, and so on. This tactical type of discussion definitely needs to occur, be addressed, and be fixed. But strategically we wanted to go to the scene, grab the hurt guy, take him to the hospital, and go back to sleep. Did we accomplish that? Maybe yes, maybe no.

The theory here is that we need to keep things in a little perspective (or in a big one). We did our jobs—ugly as it was. We must fix what went tactically wrong and feel good that strategically most of our goals are met on most of our calls. (Remember: All patients eventually die. All bleeding eventually stops. And, if you drop the baby, pick it up).

Just accomplishing our strategy is something to be pretty darn proud of by itself. Lose all the hostility and finger-pointing and recognize firefighters for the sacrifices they make in good faith. This helps dull the hurt feelings that may have been created by the last section. Firefighters are darn fine people who do a dang tough job. Make sure they are proud of that fact.

Step 5—So, what are you going to do about it? Now that we know what happened, what went wrong, and what went right, what can we do to decrease the occurrences of the wrongs, increase the occurrences of the rights, and continue to provide a great public service while not eating our young? The lessons learned from this incident should be reviewed by someone with authority to make the changes. It does no good for a junior officer to stand up and say, "We'll change our SOPs," when they have no power to do so.

Someone with authority must take center stage and review the lessons learned so everyone knows what can be expected of the department, the leaders, and the firefighters on the next incident. This wrap-up is the closure for the incident and the debriefing, and is very important. Too often we leave the room wondering if we are going to, "Do it Don's way the next time." Make the plan of implementation of suggestions known—immediate, upon review by the chief, after further investigation, etc.

That is not to say that firefighters cannot immediately make changes and adjustments in they way they operate. There are many lessons learned that do not require a two-thirds majority of Congress to enact. Many of these simple lessons can be used immediately to improve productivity, safety, and teamwork.

THE DO'S AND THE DO NOT'S

As an instructor, I find it is very easy to turn a debriefing session into an instructor-led lecture of the incident. This will kill your program and your support. As you know, adults learn best by comparing and applying their previous experiences and their existing knowledge to the current problem. The best way to ensure this happens is to let them tell their own story and discuss it. Instructors can talk until they are hoarse and maybe the information will be retained by the student and maybe not. But allow the students to analyze the problem and "discover" the solution on their own, and they will feel good about themselves and their abilities. Also, they will remember the solution in times of need and stress better than the lecture method.

The goal of an instructor during a debriefing session is to be a facilitator, a listener, a guide who has a map of where he wants the students to go, but does not show it to them—not exactly. As a facilitator, you must be ready and prepared to use open-ended questions, to target the quiet individuals, to glean information from the bearer, and to nudge discussions towards the ultimate goal—learning.

There are some calls that have gone so poorly that the group needs to be split up into officers, firefighters, and those younger than 18 years old. These critiques can be very rigorous, both emotionally and intellectually. Others are so *touchy-feely* that you expect a hug as you leave. Hopefully the facilitator can reach a happy medium between these two extremes, but I am almost sure that you will be involved in both during your career.

Here are some general guidelines to help facilitate a debriefing session:

The Do's of Facilitation

Do: Set the expectations for participation and the Blue Line.

Do: Facilitate the debriefing to achieve the objectives.

Do: Use the level and method of facilitation the group requires. Remember the key— facilitator talking = less learning; student talking = more learning.

Do: Specifically draw out quiet or reluctant firefighters.

Do: Complete all five steps of the debriefing process.

Do: Expand on important instructional points.

Do: Make the positives known and draw those out of people at the proper time.

The Do Not's of Facilitation

Do Not: Lecture as the facilitator during a debriefing. Critiques are instructor-led, but not debriefings.

Do Not: Present an analysis of the situation as a facilitator, if a firefighter can do it in his or her own words. Expand on those words if necessary, but only enough to clarify a point or to guide the group member.

Do Not: Give the impression that you have all the answers before, during, or after.

Do Not: Interrupt a speaker or discussion unless absolutely necessary.

Do Not: Abbreviate a session just because the scene went well. All team members can learn something from the incident and debriefing. It may even strengthen the team by allowing them time to talk about their jobs and actions.

SFRM[1]

NASA-Johnson Space Center is the organization that trains America's astronauts. They train shuttle astronauts, space station residents, mission controllers, and hundreds of others who are involved in the risky venture of manned space flight. Overall their safety record is phenomenal. They are constantly training for the unexpected and watching for it as well. What NASA has done in implementing their CRM program is a benchmark for all industries.

They call their program: "Spaceflight Resource Management (SFRM)." Their program is culture specific; it uses pertinent, reality-based scenarios and case studies; and they have implemented the key to a successful operational organization and information transfer. They have developed a tool, a rating instrument that assists in the operational debriefing of a mission and addresses the human behavior side of the equation. This makes their SFRM program successful, and is the key to making your Fire Service CRM Program successful, too.

By combining good facilitation techniques and a CRM Instrument to your debriefing sessions, you accomplish many necessities simultaneously.

1. You gain the advantage of transferring real-world knowledge and experience to your firefighters in a timely manner.

2. You increase the opportunities to build your team into a strong, cohesive unit whose members continually increase their knowledge of their team members.

3. You decrease the need for refresher trainings established by a time frame rather than need. (These debriefing sessions can often be productive enough to qualify for existing, mandatory, training requirements and should be listed as such.)

4. You delete the requirement for time-based refresher training in the CRM topics, because—

5. The CRM Instrument brings together operational debriefing items and human factors issues. This allows both to be discussed in one session, and not independently since significant positive or negative CRM Behavior is directly tied to successful or unsuccessful operations.

A debriefing instrument from NASA's SFRM is adapted to the fire service and included in Table 9–1.

THE WRAP-UP

Your department's culture is a complexity of history, environment, situations, resources, humans, and a hundred other things. To change any one is a serious undertaking; to change more than one requires much support and forethought. Through this process of debriefing and continuing CRM conceptual and application discussions, the culture and expectations of your department will slowly begin to change. Your personnel will begin to visualize the human factors issues involved in operations during the planning stages and apply what they learn to those problems for a successful outcome.

By integrating debriefing and CRM, we will transfer much needed knowledge and experience to the firefighters the second best way we can—by the "Lessons Learned" discussions. Of course the best way to transfer information is by actual experience. Unfortunately experience may kill our friends or us, and we do not need fatal experiences over and over again.

Use these debriefing suggestions to build better, safer firefighters.

Table 9–1
Fire Service Crew Resource Management Debriefing Instrument

Dangerous 1	Unsafe 2	Safe 3	Safe and Effective 4	Optimal 5

Rate	Category	Comments

Decision-Making

Decisions were timely?

Appropriate Decision-Making Style Utilized?

Team selected an acceptable course of action?

Alternatives considered/discussed?

Decisions monitored?

Situational Awareness (SA)

Team maintained SA?

Vigilance maintained during periods of stress?

Stress, fatigue, and complacency

recognized and addressed satisfactorily?

Critical operational tasks completed within the expected time frame?

Communications

Communications were clear and timely?

Information and actions verified?

Team asks for clarification on unclear or ambiguous communications?

Barriers/Filters addressed overcome?

Inquiry/Advocacy utilized appropriately?

Workload Management

Team aware of strategy and tactics progression?

Priority operations identified and supported?

Table 9–1 (cont.)
Fire Service Crew Resource Management Debriefing Instrument

Dangerous 1	Unsafe 2	Safe 3	Safe and Effective 4	Optimal 5
Rate	Category		Comments	

Team anticipated upcoming events/actions?

All teams/members used effectively?

Leadership

Prebriefing adequate to form a shared mental model?

Team formation addressed?

Aware of technical competency issues?

Leadership supported CRM?

Leadership consistent and appropriate?

Leadership followed applicable SOPs and requirements?

Verification of critical information and operations?

Followership

Physical needs of team members addressed?

CRM concepts supported by members?

Technical proficiency issues addressed?

Members followed applicable SOPs and requirements?

Micro-Training Opportunities utilized?

Organization

SOPs/requirements appropriate for the situation?

Adequate resource support?

Interagency cooperation?

Supported Leadership/ Followership Interfaces?

REFERENCES

1. We had the opportunity to participate in NASA's SFRM course for students and facilitators at the Kennedy Space Center in Houston, Texas. Many thanks to David G. Rogers, Brad Miller, and Steve Dillon for the priceless opportunities they provided for us. Other developers of SFRM include: Ronda Kempton, John Bauer, Derone Jenkins-Todd, Ph.D., Michael Sterling, and Therese Huning.

Special thanks to Lori McDonnell, Kimberley Jobe, Key Dismukes, and the NASA Ames Research Center for their efforts in studying the effective methods of debriefing. Their manual, "Facilitating LOS (line-oriented simulations) Debriefings: A Training Manual" has been an important part of this chapter.

STRATEGIES FOR IMPLEMENTATION

The only thing more overrated than implementing a CRM program in a fire department are the joys of natural childbirth—or so my wife says. Pitfalls abound in the implementation of a CRM program. Barriers to implementation include the areas of psychological, political, funding, administrative, and training materials.

In our experience, the most difficult barriers to overcome are the psychological barriers. Firefighters are a hands-on group of folks, who like to do "things" instead of thinking about "concepts." They like tradition and they are comfortable with the status quo. Telling a firefighter there is a new way of thinking that will make your on-scene operations more effective, is like telling a three-year-old not to eat the cookie on the counter because it will spoil dinner. The cookie is going to vanish, if it is not placed where the three-year-old cannot get it, dinner or not.

In much the same way, the firefighter realizes there is an important concept, but overcoming the inertia of, "this is the way we've always done it" is a difficult task.

Changing behavior is a difficult thing to do, and implementing a CRM program is a task akin to quitting tobacco products. It can be done, but it takes willpower and help.

Implementation of a CRM program requires rethinking of operational protocols from the top down and from the bottom up. Buy-in at the command level is essential to CRM success. The command officer must expect an information transfer from the bottom up, and should look at that information as a gift from the firefighter sending the information. Each bit of information is essential to the success of the operation. Relevance of the information can only be determined by the decision-maker.

A firefighter tells the officer from Division C that there are some strange creaking noises coming from the building (inquiry). That information may be irrelevant, or it may save the lives of the folks who are conducting the interior attack. If the firefighter then advocates a position with a statement like, "We hear some strange creaking from the building. Do you think that could be an indicator of a potential collapse?" Some officers would view that statement as insubordination. In reality, it is two minds with different information sets, trying to prevent a really bad thing from happening. Interactions between command and lower level folks will change. That change will make the department more efficient.

Political barriers serve as an impediment to implementation. In some departments that have tried to implement this program, questions start coming that the leadership does not want to answer. After an initial buy-in to the program, autocratic leaders find the requests for information unacceptable. Those leaders, who maintain their power base by hoarding the relevant information, tend to be unable to cope in a situation where operational information is transferred freely between all levels of an organization. Then the autocratic leaders begin to put the brakes on the program.

On the other hand, those leaders who freely share information, find the program to be a wonderful augmentation to their resources.

They learn that by giving the team the tools that the team needs to be effective, the program works wonders in terms of safety and efficiency of operations.

Correct implementation of a CRM program can be an expensive venture. All levels of the organization need to be trained in the underlying principles. The training is expensive in terms of time. Training materials are scarce. Those that are available come with a cost. The benefits of the program are hard to quantify and so a cost-benefit analysis is difficult. The airline industry has been grappling with the concept for years. In the same way, it is hard to quantify how many airplanes did not crash, it is hard to quantify how many firefighter deaths or injuries will be avoided if CRM programs are effectively implemented. Those airlines, as well as those fire departments that have implemented the program swear by the results. The FAA has bought into the program to the point that CRM programs are required by law for air carriers. The time may come when CRM training is mandated by Occupational Safety and Health Administration (OSHA) in the same way that HAZMAT operations are governed by CFR 1910.120.

The administrative barriers to implementation of a CRM program can be significant. Time is limited and training budgets are strained. Scheduling a comprehensive rethinking of departmental operations is a time consuming and difficult task. Planning should be done at the administrative level to implement the program, and should be phased in over a period of time.

Our experience in implementation of the program, which has met some success, consisted of the following steps. First, we began to subtly introduce CRM concepts in departmental training. We whetted the appetites of our firefighters for learning the new concepts. Next, we conducted a full-scale CRM training, consisting of all of the modules. Finally, we integrated CRM concepts into our regular training regimen. Instead of teaching CRM as a wholly separate subject, we included the concepts in our regular training, and made CRM a part of our culture.

We were fortunate in that we had departmental leaders who were willing to try something totally new and untested. Any outside person who had had contact with our department will testify to the success of our program. We routinely hear that we have something special in our department, but others can't figure out just what it is. It is a team approach to solving problems, where everyone's input is respected, but command has the final decision.

Finally the training materials are limited. At the writing of this book, two national courses are available to members of the fire service. The National Wildfire Coordinating Group has developed "Human Factors on the Fireline." The IAFC has a booklet on CRM for the fire service. Both are good basic programs that do a fine job of introducing CRM concepts to the firefighter. Additionally, there are several regional training programs available. With this text, we hope to get the message out to the rest of the firefighters in the nation. This concept has worked well for the aviation industry, and is finding its way into all operations that require decisions to be made in time-stressed situations, where there is limited information and the possibility of dire consequences. Organizations that have looked at this program include the military, the nuclear power industry, the medical industry and now, the fire service.

THE TRAINING EFFORT

After administrative approval of the program is obtained, the training effort must be carefully crafted to achieve success in the fire service. When the airline industry first began to implement this program, it got the reputation as "charm school" among the pilots who took the program. Before a training program can be effective, there must be a buy-in of all involved.

We obtained buy-in effectively, but by accident and because of an accident. We had reviewed the research, and we knew that 80% of junior officers in the airline industry would not tell command about a dangerous condition until it was too late to do anything about it. At our first formal CRM program, we asked the question to a mixed audience of probationary firefighters, experienced firefighters, and command officers, "Would you tell command if you saw a dangerous condition?" Consistent with the results of the aviation industry, we found that a large majority of our firefighters would not tell command. When firefighters were questioned about their response, they told us, "They are command. They already know."

With that statement, we had immediate buy-in from our command officers. We don't know if such a bold experiment could work in other departments, and if not, it is probably because that department is already using CRM principles without a name. We suggest that it is a question that should be asked of your department.

After obtaining buy-in, we conducted the training. Our training model is as follows:

Step 1: Train to technical firefighting proficiency

Step 2: Train to CRM proficiency

Step 3: Train risk versus gain

Step 4: Include CRM in the department culture, debriefing, and training

Step 1: Train to technical firefighting proficiency

The CRM program only works when the firefighters are trained to technical proficiency. A firefighter must know not only what to do in a given situation, but why the task is done that way. One of the reasons that CRM works for the aviation industry is that every person sitting in an airliner's cockpit has been trained for at least 1500 hours before they even take a seat in the airliner. All pilots know what they are doing and why they are doing it. They understand the importance of critical bits of information, and know what is relevant to share, and what is irrelevant.

The CRM program breaks down when a firefighter who does not know the underlying theories and what to look for, wastes time by advocating a silly position. Thus training to technical proficiency is a must.

Step 2: Train to CRM proficiency

Once the firefighter was trained to technical proficiency, then we trained the firefighter to CRM proficiency. All of the concepts contained in this manual were part of the CRM training.

Step 3: Train risk versus gain

Finally, we trained the firefighters on a risk-versus-gain analysis. The old saying goes, "Risk a lot to save a lot. Risk little to save little." Sometimes in the excitement of firefighting, we risk a lot to save a building, which is going to be torn down in a week anyway. We can't bring firefighters back from the dead. We can't unburn charred skin. We've all seen it a hundred times or more, the firefighters who are willing to risk their lives to save a couch.

We train our firefighters to decide whether what they are doing is worth a life of constant pain, or worse, the end of their lives. We take the rule seriously that every firefighter is responsible for the safety of every other firefighter. With that responsibility, we give authority. If someone asks, "What are we doing? This is stupid!" We all take a step back and figure out just what it is we are trying to accomplish. With that analysis, we have accomplished the underlying theme of CRM—to save firefighters from stupid losses.

If it is stupid to run code on a busy street and we are not going to get there much faster or in time to make a greater difference, we turn off our lights and sirens. The fun and adrenaline of running code is gone, but so are the risks of getting killed in a traffic accident. In that way, we hope to save lives.

Step 4: Include CRM in the department culture, training, and debriefing

To successfully implement this program, you cannot run a weekend program and then say, "We've implemented CRM. Now our operations will run better." The key to implementation of this program, and making it work is to include CRM in the department's culture. Taking care of equipment is part of a department's culture. The department ensures that the equipment is washed, waxed, and operational in a high state of readiness. CRM is a similar way of thinking. However instead of taking care of equipment, the department takes care of its people, and the interactions between them.

Add CRM elements to the debriefings after incidents, using the checklist from the debriefing chapter. Train the CRM elements of an operation when you do technical skills training. Ask, "When you conduct this operation, what is it you think command will want to know?" Train your firefighters on the information command needs to make an informed decision before they get into a situation where the information is needed.

Stop training operations and test firefighter situation awareness. Ask the firefighters what is going on in different segments of the operation. Ask what is being seen and heard, and why it is relevant to what is happening? Ask if there is anything happening that is out of place.

Model good communication skills off the fireground. Use the techniques outlined earlier in this book off the fireground. Train like you play.

Make leadership and followership skills a part of the day-to-day operations of the department. If you give a firefighter the responsibility, make sure you give the authority, too.

Adoption of a CRM program is not easy. You will find that you have to drag some of your firefighters kicking and screaming into the 21st century, but the fact of the matter is that's where they live, now.

If they want to be firefighters, have them be 21st century firefighters. The 19th century, with all of its traditions, has been over for a long time.

The key to successful implementation of a CRM program is constant reinforcement of CRM principles in day-to-day operation of the department. In order to be successful, CRM must become part of the culture of an organization, not just another safety program.

CONCLUSION

The road is difficult. The challenges are many. The rewards are worth the effort. Save lives. Implement a CRM program.

Index

A

Accountability, 53

Administrative requirements (team meeting), 166

Advocacy (communications), 116–119, 123:
example, 117–118;
skills, 123

After-action reviews (debriefing/critiques), 255–256

"Ah pooh" experience, 82–83

Aids (decision-making), 242–246:
checklist, 242–245;
timers, 245;
officer's aide, 246

Airshow syndrome, 199–200

Anger issues, 125–126

Anti-authority, 196

Antidotes (hazardous thoughts), 200–201

Application
(CRM leadership), 180–186:
practicals before incident, 181;
practicals during call-out, 181–182;
en route to incident, 182–183;
on-scene responsibilities, 183–184;
feedback, 184–186;
after the call, 186

Assertiveness/authority balance, 221

Assessing problems, 87–88

Assignments, 117–119, 222, 261:
followership, 222;
actions, 261

Assumptions
(communication), 108–112:

Attitude, 50–53, 80–81, 85, 108–112, 195–196, 198–199, 204–205, 218–219

Authority, 196, 214–215, 221:
anti-authority, 196;
respect for, 214–215;
balance with assertiveness, 221

Aviation industry
(CRM history), 2–3, 10, 14–16:
research application to
firefighting, 15–16

Awareness
(situational), 21, 65–95,
112–113, 247:
definition/parameters, 65–66;
training, 66–67;
loss, 68;
clues to loss, 69–83;
developing tools
to maintain, 84–85;
strategy for maintaining, 86–89;
memory, 90–94;
conclusion, 94;
references, 95–96;
communication problems,
112–113

B

Baby boomer generation, 136–137

Barriers to implementation (CRM),
33–36, 271–274:
safety culture, 33–36;
budget, 35, 273;
administrative, 271–274;
politics, 272;
information sharing, 272–273

Blue line atmosphere, 257–258:
rules, 258

Briefing (fireground), 43–44

Budget barrier, 35, 273

Bureau of Indian Affairs, 17

Bureau of Land Management,
13, 17

Busy-work tactics, 52–53

C

Change of operational setting,
62–64

Change resistance, 7–9

Checklists
(decision-making), 242–245

Chunking (memory), 91–92

Clarify order/expected action,
114–115

Classroom training, 167

Clues (loss of situational
awareness), 69–83:
fixation, 69–71;
overconfidence, 71–72;
distraction, 72–74;
information overload, 74–75;
communication, 76–77;
low stress level, 77;
high stress level, 77–78;
lack of experience, 78;
fatigue/illness, 79;
reliance on machines, 79–80;
unresolved discrepancies, 80;
professional attitude, 80–81;
looking for ghosts, 81;
gut feeling, 81;
"ah pooh" experience, 82–83

Command, Leadership, Resource
Management (CLR), 2

Communication assumptions,
108–112

Communication purposes/
objectives, 104–106:
human, 105–106

Communication skills, 216–218

Communication system, 114–124:
inquiry, 114–115;
clarify order/
expected action, 114–115;

advocacy, 116–119, 123;
Tenerife, Canary Islands
   example, 120–122;
monitoring, 123;
listening, 123–124

Communication types, 98–103:
verbal, 99–101;
written, 101;
non-verbal, 101–102;
symbolic, 102–103

Communications, 22, 76–77, 88,
97–132, 174–176, 185–186,
216–218:
effective, 76–77, 107–108,
   112–113, 128–130;
types, 98–103;
computer operations, 103–104;
why communicate, 104–106;
purposes/objectives, 104–106;
human, 105–106;
orders, 107–108;
assumptions, 108–112;
awareness of problems, 112–113;
communication system, 114–124;
filters, 124–127;
references, 130–132;
skills, 216–218

Computer communications,
103–104

Confidence/overconfidence, 71–72

Conflict resolution
(communications filters),
125–126

Confrontation with leader,
212–214:
deliberation/discretion, 213–214

Consistency (team building), 161

Constructive criticism, 209.
See also Critiques and debriefing
and Feedback.

Contingency planning.
See Situational awareness.

Control (maintaining), 86

Creating (safety culture), 36–38

Crew resource management
(CRM), 1–23, 133–188,
276–278:
human factors, 2–5, 19–21;
problem with safety
   programs, 5–9;
concept and history, 10–19;
cards we are dealt, 19–21;
chapter overview, 21–23;
references, 23;
leadership, 133–188;
proficiency, 276;
reinforcement, 277–278

Critiques and debriefing, 22–23,
209, 253–270, 277–278:
criticism, 209;
missed opportunities, 255;
after-action reviews, 255–256;
blue line atmosphere, 257–258;
pathway to enlightenment,
   258–263;
facilitation, 263–265;
Spaceflight Resource
   Management, 265–266,
   268–269;
wrap-up, 267;
references, 270;
CRM, 277–278.
See also Feedback.

CRM concept and history, 10–19:
fire service, 10–14, 16–19;
Storm King Mountain/
   South Canyon fire
   investigation, 10–14;
aviation industry, 14–15;
aviation research application
   to firefighting, 15–16;
status in fire service, 16–19

CRM leadership, 133–188:
  introduction step, 138;
  integration step, 139–141;
  trust step, 141–143;
  who is the leader, 143–149;
  leader activities, 150;
  lack, 151–152;
  team building, 153, 156–161;
  team performance issues,
    153–156;
  organization's role in team
    building, 161–165;
  leader's responsibilities
    (teams), 165–176;
  practical leadership, 177–180;
  application, 180–186;
  conclusion, 187;
  references, 187–188
CRM proficiency, 276
CRM reinforcement, 277–278
Cultural norms
  (followership), 208–209
Culture change, 27–31:
  fire service is paramilitary
    organization, 27–28;
  organizational change, 28–31

D

Debriefing (decision-making),
  237–238
Debriefing and critiques, 22–23,
  237–238, 253–270, 277–278:
  decision-making, 237–238;
  missed opportunities, 255;
  after-action reviews, 255–256;
  blue line atmosphere, 257–258;
  pathway to enlightenment,
    258–263;
  concepts, 259–260;
  template, 260–263;

facilitation, 263–265;
  Spaceflight Resource
    Management, 265–266,
    268–269;
  wrap-up, 267;
  references, 270;
  CRM, 277–278
Debriefing/critiques concepts,
  259–260:
  timing, 259–260;
  everyone involved, 260;
  length, 260
Debriefing/critiques template,
  260–263:
  incident facts, 260;
  assignments and actions, 261;
  what went wrong, 261–262;
  what went right, 262–263;
  lessons learned, 263
Decision models, 230–233
Decision-making, 13–14, 22,
  229–251:
  decision models, 230–233;
  substitute for experience, 233;
  improvement, 234–242;
  debriefing, 237–238;
  preplan, 238–240;
  aids, 242–246;
  environment, 246–247;
  steps, 247–248;
  evaluation, 248;
  on-scene tips, 249;
  conclusion, 250;
  references, 250–251
Deliberation/discretion
  (confrontation), 213–214
Direction (team building), 160
Direction/information acceptance,
  222
Discrepancies unresolved, 80
Distraction, 72–74

E

Effective communications, 76–77, 107–108, 112–113, 128–130

Ego in check, 219–221

Emergency calls, 82–83

Emergency medical services, 4, 50

Emotional attitude (followership), 204–205

Emotional/physical condition, 85

Enlightenment (debriefing/critiques), 258–263: concepts, 259–260; template, 260–263

Entrapment causes, 11: fire behavior, 11

Environment (decision-making), 246–247

Equipment management, 13

Erroneous assumptions (communication), 108–112

Error-free world, 31–32

Errors/omissions (followership), 223

Evaluation (decision-making), 248

Expectations, 18–19, 45–46, 166–167, 192: fireground, 45–46; team meeting, 166–167; self, 192

Experience, 48–50, 78, 84, 233: as training, 48–50; lack, 78; situational awareness, 84; decision-making, 233

Extemporaneous teams, 173–176

F

Facilitation (debriefing/critiques), 263–265: do's and do not's, 264–265

Fatal decisions, 13–14

Fatalities (firefighter), 4–6, 13–14, 16–17, 28

Fatigue, 79, 194

Federal Aviation Administration (FAA), 2, 15, 97

Feedback, 126, 184–186, 209: communications filters, 126; incident, 184–186; criticism, 209. See also Critiques and debriefing.

Filters (communications), 124–127: conflict resolution, 125–126; feedback, 126; system in use, 127

Fire behavior, 11, 177–178: fuels, 11; weather, 11; topography, 11; predicted behavior, 11; observed behavior, 11

Fire service (CRM history), 10–14: Storm King Mountain/ South Canyon fire investigation, 10–14

Firefighting skills, 85, 169

Fireground operations formula, 41–47: briefing, 43–44; desired results, 44–45; expectations, 45–46; responsibilities, 46–47; available resources, 47

Fixation, 69–71

Follower role, 107–108, 205–207, 209–212, 214, 216–218:
orders, 107–108;
training, 207;
helping the leader, 209–212;
skills, 214;
communications, 216–218

Followership, 22, 33–34, 107–108, 136–137, 189–227:
orders, 107–108;
definition, 190–191;
junior personnel tendencies, 191–201;
hurry-up syndrome, 201–203;
recommendations, 203–223;
role, 205–207;
training, 207;
helping the leader lead, 209–212;
skills, 214;
communications, 216–218;
conclusion, 224;
references, 225–227

Fuels (fireground), 11

G

Generation X, 137

Ghosts (fireground), 81

Goal of operations, 30, 44–45, 158–159

Gut feeling (fireground), 81

H

Hazardous materials (HAZMAT) checklist, 244

Hazardous thoughts, 196–201:
anti-authority, 196;
impulsivity, 196;
invulnerability, 197–198;
macho attitude, 198;
resignation, 198–199;
pressing, 199;
airshow syndrome, 199–200;
antidotes, 200–201

Helping the leader lead, 209–212

High stress level, 77–78

Human communications, 105–106:
realities, 106

Human error, vii–viii, 2–5, 7–9, 13–14, 19–21, 31–32, 36–38, 98, 108–112, 223:
Swiss cheese error model, 20–21;
erroneous assumptions, 108–112

Human factors, 2–3, 7–9, 13–14, 17–21, 32–33, 98, 65–95, 106, 112–113, 247:
risk homeostasis, 7–9

Hurry-up syndrome, 201–204

Hydration, 195

I

Illness/fatigue, 79

Implementation barriers (safety culture), 33–36

Implementation strategies (CRM), 23, 30–31, 271–278:
training effort, 274–278;
conclusion, 278

Impulsivity, 196

Incident facts (debriefing/critiques), 260

Incident management (South Canyon fire), 12–13:
strategy and tactics, 12;
safety briefing and concerns, 12;

involved personnel profile, 12; equipment, 13

Incident practicals, 181–186: before incident, 181; during call-out, 181–182; en route to incident, 182–183; on-scene responsibilities, 183–184; feedback, 184–186; after the call, 186

Increasing memory (strategies), 91–94: chunking, 91–92; visual echoes, 92; over-learning, 92–94

Individual responsibility, 32–33

Informal leader, 144–147: know team individuals, 146; create mini-opportunities, 146–147; support, 147; performance analysis, 147

Information, 9, 41–47, 74–75, 85, 88, 222, 272–273: overload, 74–75; processing, 85; gathering, 88; acceptance, 222

Initial meeting (team), 165–167: preparation, 166; introduction, 166; statement to group, 166; administrative requirements, 166; expectations, 166–167

Inquiry (communications), 114–115: example, 115; skills, 115

Integration (CRM leadership), 139–141

International Association of Fire Chiefs, 17

International Association of Fire Fighters, 17

Introduction (CRM leadership), 138

Introduction (team meeting), 166

Involvement (debriefing/critiques), 260

Invulnerability, 197–198

J–K

Junior personnel tendencies, 191–201

L

LCES checklist, 243

Leader confrontation, 212–214: deliberation/discretion, 213–214

Leader identification, 143–149: informal leader, 144–147; situational leader, 148–149

Leader responsibilities (teams), 165–176: initial meeting, 165–167; new team, 167–170; old team, 170–173; extemporaneous teams, 173–176

Leadership (CRM), 22, 33–34, 133–188, 209–214: introduction step, 138; integration step, 139–141; trust step, 141–143; who is the leader, 143–149; leader activities, 150;

lack, 151–152;
team building, 153, 156–161;
team performance issues,
   153–156;
organization's role in team
   building, 161–165;
leader responsibilities
   (teams), 165–176;
issues, 171;
practical leadership
   performance, 177–180;
application, 180–186;
conclusion, 187;
references, 187–188;
helping, 209–212;
confrontation, 212–214

Leadership performance
   (CRM), 177–180:
what if's, 178–179;
technical proficiency, 179–180

Learning attitude, 218–219

Lessons learned
   (debriefing/critiques), 61–62,
   181–182, 263

Listening
   (communications), 123–124

Loss (situational awareness), 68–83:
clues to loss, 69–83

Low stress level, 77

M

Machine reliance, 79–80

Macho attitude, 198

Maintaining situational awareness,
   84–89:
developing tools, 84–85;
experience, 84;
training, 84;

firefighting skills, 85;
ability to process information, 85;
professional attitude, 85;
emotional/physical condition, 85;
strategy, 86–89;
maintain control, 86;
assess problem in
   time available, 87–88;
gather information, 88;
monitor results, 89

Media age, 137

Memory
   (situational awareness), 90–94:
effects of stress, 90–91;
strategies for increasing, 91–94;
chunking, 91–92;
visual echoes, 92;
over-learning, 92–94

Mental attitude, 195–196, 198

Mental models
   (team building), 160

Micro-training opportunities, 48–50

Missed opportunities
   (debriefing/critiques), 255

Mission analysis
   and planning, 21, 39–64:
formula for fireground
   operations, 41–47;
micro-training opportunities,
   48–50;
positive/proactive attitude,
   50–53;
accountability, 53;
risk of operational activity, 53–57;
risk vs. gain analysis, 58;
risk acceptance, 58–62;
changing operational setting,
   62–64;
references, 64

Model (error), 20–21

Monitoring, 89, 123:
  results, 89;
  communications, 123

N

National Aeronautics and Space
  Administration (NASA), 2, 15,
  43, 265–266, 268–269
National Association of
  State Foresters, 17
National Park Service, 17
National Transportation
  Safety Board (NTSB), 2, 15
National Wildfire
  Coordinating Group, 17
New team, 167–170:
  classroom training, 167;
  scenario tabletops, 168;
  physical training, 168–169;
  skills training, 169;
  reality incidents, 169;
  unstructured time, 169–170
Non-verbal communications,
  101–102
Nutrition, 194–195

O

Observed behavior, 11
Officer's aide
  (decision-making), 246
Old team, 170–173:
  leadership issues, 171;
  personal issues, 171–172;
  organizational support, 172–173;
  self factors, 173

On-scene tips
  (decision-making), 249
Operational activity risk, 53–57:
  structural firefighting, 56–57;
  wildland firefighting, 56–57
Operational setting
  (change), 62–64
Operations formula
  (fireground), 41–47:
  briefing, 43–44;
  desired results, 44–45;
  expectations, 45–46;
  responsibilities, 46–47;
  available resources, 47
Operations goal, 30, 44–45,
  158–159
Orders made to be followed,
  107–108
Organization's role
  (team building), 161–165
Organizational culture
  change, 27–31:
  fire service is paramilitary
  organization, 27–28;
  organizational change, 28–31
Organizational safety culture, 21,
  25–38, 53–57, 277–278:
  culture change, 27–31;
  error-free world, 31–32;
  individual responsibility, 32–33;
  barriers to implementation,
  33–36;
  creating, 36–38;
  references, 38
Organizational support, 172–173
Overconfidence, 71–72
Over-learning (memory), 92–94

P–Q

Paramilitary organization
(fire service), 27–28

Pattern creation/recognition,
232–233

Performance analysis, 147

Performance issues, 153–156,
177–180:
team, 153–156;
leader, 177–180

Personal Accountability
Safety System (PASS), 93

Personal issues, 171–172

Personnel, 12, 191–201:
profile, 12;
junior personnel tendencies,
191–201

Physical fitness, 85, 168–169,
193–194:
training, 168–169

Planning (mission analysis), 21,
39–64:
formula for fireground
operations, 41–47;
micro-training opportunities,
48–50;
positive/proactive attitude,
50–53;
accountability, 53;
risk of operational activity, 53–57;
risk vs. gain analysis, 58;
risk acceptance, 58–62;
changing operational setting,
62–64;
references, 64

Positive/proactive attitude, 50–53

Practical leadership (CRM),
177–180

Practicals, 181–184:
before incident, 181;
during call-out, 181–182;
en route to incident, 182–183;
on-scene responsibilities, 183–184

Predicted behavior, 11

Preparation
(team meeting), 166

Preplan
(decision-making), 238–240

Pressing attitude, 199

Problem assessment, 87–88

Professionalism, 80–81, 85, 154

Protective clothing, 3, 7–8, 13, 16

R

Real life training, 48–50

Realities
(human communications), 106

Reality incidents
(training), 169

Reason model, 20

Recommendations
(followership), 203–223:
emotional attitude, 204–205;
role of follower, 205–207;
training, 207;
safety, 207–208;
cultural norms, 208–209;
constructive criticism, 209;
helping the leader lead, 209–212;
leader confrontation, 212–213;
deliberation/discretion, 213–214;
skills, 214, 216–218;
respect for authority, 214–215;
safety eyes, 215;
technical competence, 216;
communication skills, 216–218;

learning attitude, 218–219;
ego in check, 219–221;
authority/assertiveness
balance, 221;
accepts direction and
information, 222;
demands clear assignments, 222;
admits errors/omissions, 223

Reliance on machines, 79–80

Resignation, 198–199

Resistance to change, 7–9

Resources available
(fireground), 47, 248

Respect for authority, 214–215

Responsibilities
(leader/team), 165–176:
initial meeting, 165–167;
new team, 167–170;
old team, 170–173;
extemporaneous teams, 173–176

Responsibility, 32–33, 46–47,
165–176, 183–184:
individual, 32–33;
fireground, 46–47;
leader/team, 165–176;
on-scene, 183–184

Results desired
(fireground), 44–45

Results monitoring, 89

Rewards (team building), 161

Risk acceptance, 58–62:
possible gains, 60;
possible risks, 60

Risk homeostasis, 7–9

Risk of operational activity, 53–57:
structural firefighting, 56–57;
wildland firefighting, 56–57

Risk vs. gain, 58, 276:
analysis, 58

Risk/risk analysis, 7–9, 31–32,
53–62, 276:
homeostasis, 7–9;
operational activity, 53–57;
structural firefighting, 56–57;
wildland firefighting, 56–57;
risk acceptance, 58–62;
risk vs. gain analysis, 58, 276

S

SAFE STOP checklist, 244–245

Safety
(followership), 207–208, 215

Safety briefing and concerns, 12

Safety culture, 5–9, 12, 21, 25–38,
53–57, 207–208, 215, 277–278:
problem, 5–9;
briefing/concerns, 12;
culture change, 27–31;
error-free world, 31–32;
individual responsibility, 32–33;
barriers to implementation,
33–36;
creating, 36–38;
references, 38;
followership, 207–208, 215;
safety eyes, 215

Safety eyes
(followership), 215

Safety infractions, 6

Safety programs, 5–9, 25–26,
28–29:
problems, 5–9;
infractions, 6

SAMPLE checklist, 243

Scenario tabletops
(training), 168

Self factors, 173

Self-contained breathing apparatus (SCBA), 3, 16

Self-expectations, 192

Shared mental models (team building), 160

SIN checklist, 244

Situational awareness, 21, 65–95, 112–113, 247:
definition/parameters, 65–66;
training, 66–67;
loss, 68;
clues to loss, 69–83;
developing tools to maintain, 84–85;
strategy for maintaining, 86–89;
memory, 90–94;
conclusion, 94;
references, 95–96;
communication problems, 112–113

Situational leader, 148–149

Skills (followership), 207, 214, 216–218

Skills training, 169

South Canyon fire investigation, 10–14:
entrapment causes, 11;
incident management, 12–13;
fatal decisions, 13–14

Spaceflight Resource Management (SFRM), 265–266, 268–269

Standard operating procedures (SOP), 6, 12, 73–74

Statement (team meeting), 166

Status quo (risk), 7–9

Steps (decision-making), 247–248:
situational awareness, 247;
technical proficiency, 247;
know resources available, 248;
evaluation, 248

Stock phrase, 100

Storm King Mountain/South Canyon fire investigation, 10–14:
entrapment causes, 11;
incident management, 12–13;
fatal decisions, 13–14

Storytelling (decision-making), 233, 240–242

Strategy (CRM implementation), 23, 30–31, 271–278:
training effort, 274–278;
conclusion, 278

Strategy (increase memory), 91–94:
chunking, 91–92;
visual echoes, 92;
over-learning, 92–94

Strategy (maintain situational awareness), 86–89:
maintain control, 86;
assess problem in time available, 87–88;
gather information, 88;
monitor results, 89

Strategy/tactics (incident management), 12

Stress effects on memory, 90–91

Stress level, 74–78, 90–91:
effects, 74–78, 90–91;
low, 77;
high, 77–78;
memory, 90–91

Structural firefighting (risk analysis), 29–30, 56–57

Swiss cheese error model, 20–21

Symbolic communications, 102–103

System in use (communications filters), 126

T

Tactical Decision Making Under Stress (TADMUS) program, 236

Tactics/strategy (incident management), 12

Team building, 153, 156–165:
shared mental models, 160;
clear direction, 160;
consistency, 161;
rewards, 161;
organization's role, 161–165

Team meetings, 165–176:
initial, 165–167;
new team, 167–170;
old team, 170–173;
extemporaneous teams, 173–176

Team performance issues, 153–156

Teams, 153–165, 167–176:
performance issues, 153–156;
team building, 153, 156–165;
meetings, 165–176;
new team, 167–170;
old team, 170–173;
extemporaneous teams, 173–176

Technical proficiency, 179–180, 216, 247, 275–276:
leadership, 179–180;
followership, 216;
training, 275–276

Technical rescue, 4

Technical-based solutions (safety), 5–6

Ten Standard Fire Orders, 140–141

Tendencies (junior personnel), 191–201:
self-expectations, 192;
physically fit, 193–194;
fatigue, 194;
nutrition, 194–195;
hydration, 195;
mental attitude, 195–196;
anti-authority, 196;
impulsivity, 196;
invulnerability, 197–198;
macho attitude, 198;
resignation, 198–199;
pressing, 199;
airshow syndrome, 199–200;
antidotes, 200–201

Tenerife, Canary Islands (communications), 120–122

Time factor, 87–88

Timers (decision-making), 245

Timing (debriefing/critiques), 259–260

Tools development (situational awareness), 84–85:
experience, 84;
training, 84;
firefighting skills, 85;
ability to process information, 85;
professional attitude, 85;
emotional/physical condition, 85

Topography (incident management), 11

Training (CRM implementation strategies), 274–278:
technical proficiency, 275–276;
CRM proficiency, 276;
risk vs. gain, 276;
CRM in department culture/training/debriefing, 277–278

Training (new team), 167–170:
classroom, 167;
scenario tabletops, 168;
physical, 168–169;
skills, 169;
reality incidents, 169;
unstructured time, 169–170

Training, 3, 17–18, 37–38, 39–41, 48–50, 66–67, 83–84, 92–94, 167–170, 207, 234–236, 253–254, 274–278:
human factors, 3;
opportunities, 48–50;
situational awareness, 66–67, 84;
new team, 167–170;
followership, 207;
decision-making, 234–236;
CRM implementation strategies, 274–278
Tri-Data Study, 17, 26, 74
Triggers (operations), 123
Tripod strategy, 18
Trust (CRM leadership), 36, 141–143
Tunnel vision, 69–71

U

U.S. Fire Administration (USFA), 4
U.S. Fish and Wildlife Service, 17
U.S. Forest Service, 17
United Airlines, 2
Unstructured time (training), 169–170

V

Verbal communications, 99–101
Visual echoes (memory), 92

W–Z

Watch-out situations, 12
Weather (incident management), 11
What if's (leadership), 178–179
What went right, 262–263
What went wrong, 261–262
Wildland Firefighter Safety Awareness Study (Tri-Data Study), 17, 26, 74
Wildland firefighting (risk analysis), 56–57
Written communications, 101